At the Fringes of Science

AT THE FRINGES OF SCIENCE

MICHAEL W. FRIEDLANDER

With a New Epilogue

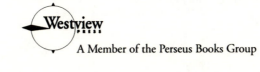

Westview PRESS

A Member of the Perseus Books Group

Copyright © 1998 by Westview Press, A Member of Perseus Books Group

Published in 1998 in the United States of America by Westview Press, 5500 Central Avenue, Boulder, Colorado 80301-2877, and in the United Kingdom by Westview Press, 12 Hid's Copse Road, Cumnor Hill, Oxford OX2 9JJ

Library of Congress Cataloging-in-Publication Data
Friedlander, Michael W.
  At the fringes of science / Michael W. Friedlander.
    p. cm.
  Includes bibliographical references and index.
  ISBN 0-8133-2200-6 (hc) — ISBN 0-8133-9060-5 (pbk).
  1. Discoveries in science.   2. Serendipity in science.   3. Fraud.
in science.   I. Title.
Q180.55.D57F75   1995
500—dc20   94-35048
                                                                  CIP

The paper used in this publication meets the requirements of the American National Standard for Permanence of Paper for Printed Library Materials Z39.48-1984.

10   9   8   7   6   5

for Jessica

Contents

Preface

Scientific discoveries are often front-page news. Whether it is the identification of the gene responsible for some hereditary disease, the observation of a gigantic stellar explosion as a supernova, or a further decrease in the stratospheric ozone, science retains an aura of the magic with which it was so long associated. For the nonscientist, that aura often has two parts: a sense of wonderment combined with a sense of incomprehension. Some discoveries may later be applied to alleviate suffering; others will remain far removed from any application, markers along the way toward a deeper understanding of our universe and its wonderful complexity.

Surrounding this domain of science there is a fringe, a penumbral zone with no sharp boundaries, filled with claimants for recognition as parts of science. In this border zone are honest errors and sloppy works; instances of fraud; "discoveries" that have been neither confirmed nor rejected. There are "scientific" determinations that represent political fiat based on expediency or dogma. Close to the region of accepted science are observations and new theories that are the current products of science, subjects of continuing research that will lead to the acceptance of some and the rejection of others. Adjacent to this region and overlapping it in places is what Irving Langmuir[1] has termed "pathological science, the science of things that aren't so," "discoveries" that may result from careless experimenting and theories based on the selective assembly of some data and the careful ignoring of others. Somewhat further from the core is an outer region, the home of pseudoscience—observations and theories that have been made to resemble science, at least in the eyes of their originators and the nonexpert. Science always has this fringe—uneven, sometimes entertaining, sometimes irritating, but overall illuminating in challenging us to clarify just what constitutes science without achieving a clear and final definition of science and its methods. The examination of these regions and their boundaries is the focus of this book.

Pseudoscience is a permanent companion to science. It survives in part because there is a large reservoir of people for whom the methods and much of the content of science remain unknown, hazy, or confused. Another reason for the attractiveness of many of the pseudosciences is that they claim to provide answers to puzzles that science has not solved. Some pseudosciences resonate with deeply felt yearnings. Parapsychology, with its early ties to spiritualism, is still in this category, along with astrology. Regrettably, some pseudoscientific "discoveries" have enjoyed their successes because they claim to provide cures where conventional medicine has failed. All of these and many other examples can be classified as pseudoscientific: They usually present the outward appearance of science but fail to sustain the illusion upon closer examination. Other episodes are not pseudoscientific but are not easy to classify. Some do not survive attempts at replication; others may be validated only after many years during which they may be considered mistaken.

Scientific error, now a subject of much comment and congressional investigation, is not encountered as often as is sometimes alleged. But it is more frequent than we would have thought, and it does require examination. Sometimes a reported "discovery" turns out to be a mistake, neither the result of fraud nor the product of ignorance. Error is a normal part of science, especially at the forefront of research, where experimental methods are being pushed to their limits. In this book I bring in some case studies to illustrate this aspect and to fill out our catalog of the fauna in the penumbral region. It is not a simple task to show nonscientists just how and where some new idea falls short of what we demand of mainstream science. We need a description of the mode of operation of conventional science. Taking this conventional model for comparison and with some appreciation of the views of the structure of science (as seen by working scientists), we give a critical survey of the penumbra some intellectual anchors.

Part of the reason for the public's widespread difficulty in separating science from nonscience is that the scientist is generally remote as a person, even though science plays so large a role in our lives. Few names and fewer faces of scientists are widely recognizable. Grade-schoolers do not collect trading cards featuring scientists and their discoveries. Pictures of scientists do not appear on cereal boxes. Scientists are too often thought of as white-coated foreigners with thick accents and thicker spectacles.

Somehow we often fail to remember that scientists have a human side. Scientists buy Girl Scout cookies, go to Little League baseball games, and complain about income tax just as so many other people do. Part of the normal range of human characteristics shows in the style of their work: Some leap to conclusions, offering hasty judgments they must later reverse; others react cautiously when they are confronted by the latest discoveries and theories; still others bend the truth. But the remarkable fact remains that, despite human fallibility, our scientific enterprise is enormously successful. Scientific opinions *are* often correct, and scientific evaluations are correct more often than not. Yet the nonscientific public seems at the same time hesitant to trust scientific opinion but willing to accept unconventional and pseudoscientific challenges to scientific orthodoxy.

My aim, therefore, is to show how scientific progress is made as new ideas emerge and are scrutinized. In doing this, I hope to demonstrate just how varied is the neighborhood surrounding science. Whatever our view of this great enterprise, we cannot escape it, and we will be better served by having an understanding of science and its methods, its imitators, and its limitations.

After taking an introductory chapter to set out the problem, I have chosen to start our survey (in Chapter 2) with two prime examples that are very different from one another. The first is the classic case of Immanuel Velikovsky and his astronomical ideas; the other is a fraudulent claim for the discovery of a compound to extend the life of automobile batteries. The differences between these two cases illustrate some of the variety in shady science. A feature common to these two cases is their occurrence outside the mainstream of science; one claimed scholarly content and received endorsement from many academics, but the other simply had a commercial agenda that was advanced through political connections. In contrast, the examples I describe in Chapter 3 (continental drift and cold fusion) attracted the serious attention of scientists within the mainstream; though one was later validated, the other seems to have been the product of optimism and sloppy techniques.

I devote the next two chapters to a description of the day-to-day workings of science, so that we will have a framework against which we can later understand the very different receptions that await new scientific or "scientific" ideas. It is often asserted that scientists do not welcome innovation; in Chapter 6 I argue against this view by describing a number of revolutionary ideas that scientists have taken very seriously.

One major feature that usually (but not invariably) distinguishes genuine science from pseudoscience is the involvement of scientists. The two examples in Chapter 7 illustrate how competent scientists can be fooled and novel ideas rejected with good reason. In contrast, the examples in Chapter 8 are the spice of the weekly tabloids, though they are often promoted in book form as well. Here we find the unidentified flying objects (UFOs), astrology, and related curiosa. I have also included a wonderful example of pseudoscience that was being created even as we watched: Iben Browning's "prediction" of a major earthquake in Missouri for December 1990 shows just how seriously a pseudoscientific claim can be taken and how much potential there is for unintended mischief. Extrasensory perception (ESP) and related phenomena deserve a chapter to themselves, for they form a distinct category of fringe science. Despite more than a century of revealed fraud and the absence of results that can convince the skeptics, this subject continues to attract the dedicated research of a few competent scientists convinced that there are genuine effects awaiting recognition by mainstream science.

Fraud, unfortunately, has turned up with increasing frequency in recent years in what had been considered clean areas of science; I explore this worrying phenomenon in Chapter 10. Pseudoscience can arrive from a different direction, gaining support from a political agenda, where scientific "fact" is asserted without a firm base. This, too, can produce distortions in science, and I describe three examples in Chapter 11. The final two chapters are devoted to an analysis of the various different examples we have covered, with the underlying question: How do and should the scientific community and general public respond to each new claim for discovery?

Perhaps I should also make clear what I am *not* trying to do in this book. I do not have the space to describe each of my chosen cases in the detail that they have received elsewhere. I provide sufficient description so that their main features can be discerned, allowing me to classify them and add my comments. There are many books devoted to critiques of astrology, UFOs, and parapsychology; I have listed a representative selection in the Further Readings. Each of these topics has a far greater literature produced by supporters and believers. As my own approach is distinctly skeptical, the bibliography reflects this, but the supporting literature can be located through my references.

I hope that this survey will be helpful in pointing to the complexity of the problem of demarcating between science, pseudoscience, and the rest. The "science" encountered in astrology, UFOs, and ESP is unlikely to help us evaluate the claims for cancer cures, understand the technical aspects of treaties for nuclear weapons, or deal with proposed environmental policies such as the disposal of radioactive waste. And unless there is a much wider and better appreciation of the strengths and limitations of science and of the distinction between science and nonscience, we will be poorly placed to handle the many problems that are scientific at base.

Michael W. Friedlander

Notes

1. *Physics Today* 42 (October 1989), 36.

Acknowledgments

It is my pleasure to acknowledge the assistance I have received. This book has grown out of seminars I have given, and I am grateful to my students, in particular those in the Master of Liberal Arts program at Washington University. A number of people—Carl Bender, Howard Boyer, Leroy Ellenberger, Leonard Green, Peter Phillips, and Douglas Wiens—have read parts of the manuscript; their comments have been especially useful. Special thanks go to my editors at Westview Press: Scott Horst, project editor, and Alice Colwell, who has greatly improved this book through her fine copy editing. And, finally, I thank my family, whose continued interest and support I treasure.

M.W.F.

I

Science and Its Imitators

"Taming H-Bombs?" was the column head that the *Wall Street Journal* supplied when it carried the first published report of the claimed discovery of cold fusion in March 1989.[1] The previous day, at a press conference at the University of Utah, Martin Fleischmann and Stanley Pons, two chemists, had announced that they had carried out experiments in which they had achieved hydrogen fusion at room temperatures.

This discovery was potentially revolutionary, holding out the prospect for an almost limitless supply of environmentally clean energy, in addition to fame and wealth. Fusion of hydrogen nuclei is the source of energy in the sun and many stars, but there it occurs at temperatures above 10 million degrees. At such high temperatures, the hydrogen nuclei move so rapidly that they can overcome the repulsion of their electric force and get close enough for the stronger nuclear force to take over. The result is fusion, the formation of a heavier nucleus. A complex chain of nuclear reactions transforms hydrogen into helium with the release of large amounts of energy accompanied by neutrons and gamma rays (more energetic versions of X-rays). A critical ingredient is the initial high temperature.

Accordingly, in their continued search for methods of fusing hydrogen under laboratory conditions for a controllable supply of energy, scientists have for many years concentrated their efforts on high-temperature processes. Room-temperature fusion, if proven, would be truly revolutionary. One exotic form of low-temperature fusion is already known, but it is mediated by muons, radioactive particles produced in high-energy nuclear collisions. And that is not what the Utah chemists were seeing.

Sadly (because of our hopes) but not surprisingly (because of our earlier experiences), after months of independent, costly, and exhaustive checks by hundreds of scientists around the world, the excitement over cold fusion cooled off, and the claim is probably destined to take its place alongside monopoles, N-rays, polywater, and other fly-by-night "discoveries" that flash across our scientific skies to end up as part of our folklore. That folklore includes a remarkable range of claims, optimistically announced, sometimes with the initial endorsement of noted scientists. Independent and more critical scrutiny usually illuminates weaknesses in the experimental procedures or mistaken assumptions, outright errors, or in a few cases (and regrettably) fraud.

Sometimes the announcements are more apocalyptic than was the case with the Utah experiment. "It may well turn out to be as epochal as *The Origin of Species* by Darwin or the *Principia* of Newton," was Clifton Fadiman's evaluation of Immanuel Velikovsky's *Worlds in Collision* when it appeared in 1950.[2] Based on his psychoanalytic analysis of the legends and epics of a wide range of peoples, Velikovsky had put forward an unorthodox theory of the behavior of the planets during the second and first millennia B.C. He had also proposed radical changes in the accepted chronology of events in the ancient world. Fadiman, a respected journalist but no scientist, was not alone in his praise of Velikovsky's book. Glowing reviews (some preceding actual publication) appeared in *Reader's Digest, Atlantic,* and *Harper's.* Eric Larrabee, *Harper's* music critic, praised Velikovsky for undertaking "an awesome task of making an inquiry in the architectonics of the world and its history."[3] Fulton Oursler, religion editor of *Reader's Digest,* said that *Worlds in Collision* "opens up a vast new debate"[4]—which it certainly did, but not in the way that Oursler had believed it would.

In another example of a well-staged presentation, the Drake Hotel in Chicago was the scene of a 1951 press conference to announce the discovery of Krebiozen, which the 106-page brochure distributed on that occasion called "an agent for the treatment of malignant tumors." The booklet mentioned a "secret ingredient" and made sweeping claims for the drug's effectiveness as a cure for cancer. Krebiozen had been developed by Dr. Stevan Durovic, a Yugoslav physician, during the years that he worked in Argentina. Now in the United States, and present at the press conference, Durovic was supported by Dr. Andrew C. Ivy, at that time distinguished professor of physiology at the University of Illinois. After its dramatic announcement, Krebiozen never showed the claimed effectiveness and faded from view, though not without some protracted court cases. Its launch and decline have been chronicled by Patricia Spain Ward.[5]

We have become accustomed to reading extravagant claims like these on the front pages of the tabloids at our supermarket checkout counter, but who takes them seriously? Scientists, in contrast, expect major discoveries to be announced through professional journals and in sober prose. So, for example, Francis Crick and James Watson's famous 1953 article in *Nature* describing their discovery of the structure of the DNA molecule is now recognized as one of the most important scientific papers of the century. Their report carried the solemn title "Molecular Structure of Nucleic Acids" and began, "We wish to suggest a structure for the salt deoxyribose nucleic acid (D.N.A.). This structure has novel features which are of considerable biological interest."[6]

In their different ways, the claims for cold fusion, for Velikovsky's theory, and for Krebiozen illustrate the gray area that surrounds the science of popular illusion, the science of brilliant discoveries and instant recognition from other scientists and a grateful public eager for the revelations that lead to medical miracles or sweeping syntheses and insights. What these examples also illustrate is the widespread misunderstanding of the way in which scientific progress arrives and in which scientific *facts* are presented and established. This confusion has been demonstrated by scientists as well as by nonscientists, and by entrepreneurs and legislators who are often influenced by nonscientific preferences.

This gray zone is the focus of this book. It covers the ways in which claimed advances, modest or revolutionary, are put forward, examined, and accepted or rejected. Some claimed advances become parts of scientific knowledge, but some are *pseudoscientific:* They appear to be scientific, make assertions that they are scientific, but on closer examination turn out to be fatally flawed in content, in method, or in both. Between outright error and obvious correctness is an untidy arena—but even the "obvious" nature of something does not become obvious at the same speed to all people. What is obvious error to a physicist or astronomer was by no means clearly so to Velikovsky and his supporters. What was obvious to critics of Pons and Fleischmann was brushed aside as quibbles by officials from the University of Utah who sped to Washington to demand $25 million from Congress to fund a new institute for cold fusion research.

How does science distinguish between the potentially genuine and the probably pseudoscientific? Examples can surely be found that will not fall neatly into either category. Faced with this fuzziness, one might be tempted to ask whether any classification at all is necessary or even useful. Instead, surely *all* new ideas should be treated as potentially fruitful and carefully examined. Science should be open, willing to embrace new ideas no matter what their origin, for only in this way will progress be made.

As is so often the case, reality intrudes. It is simply not practicable to give serious attention to *every* new idea. Many scientists receive unsolicited mail and phone calls from nonscientists who have (so they think) discovered or developed a new theory that shows the error of accepted ideas, and in physics this usually means relativity or quantum theory. I have on my shelf complete books, produced at considerable expense and distributed in the thousands, that are sad testimony to their authors' lack of understanding. Relativity and quantum theory offend the common sense of many people—sometimes physicists, too—and are the major targets of pseudophysicists. We do not find comparable rearguard guerrilla campaigns against James Clerk Maxwell, Isaac Newton, and others whose work may be less open to popular misconception. The second law of thermodynamics is the only other physics topic that draws fire, though inventors of perpetual motion machines often do not realize that this is what they are challenging. Evolution, of course, has been the target of sustained attacks for many years. I have more to say about that later.

The flaky nature of new theories of relativity or quantum effects or perpetual motion machines is quickly apparent. To take them seriously would take time away from other, scientifically more useful tasks. Experience shows that the proponents of these monotonously unconventional ideas can never be persuaded of their errors. Some time ago, a colleague and I were engaged by a would-be relativist. After I had written several letters setting out in great detail the error in his reasoning and giving him the correct approach, I had to terminate our correspondence by pointing out that we had gone around the loop several times, that the weight of professional opinion was against him, and that although he was free to believe whatever he wished, I would not respond to further letters.

Most scientists generally refuse to be drawn into this sort of discussion. It is almost certain to be unrewarding, a waste of time, though one hesitates to be totally silent and give offense. There is no reason to treat such misguided people rudely. There is also the terrible example of a deranged crank who, unable to get his manuscripts published, invaded the offices of the American Physical Society and shot a secretary. And one of my own frustrated correspondents predicted, "If I can't get you in this world, I'll get you in the next."

Challenges to accepted scientific ideas can come from many directions. The issues raised can be equally diverse and often unpredictable. The initial reception accorded new ideas depends almost entirely on the credentials of their originators. Known scientists are more likely to be taken seriously, as were Pons and Fleischmann and their claims for the demonstration of cold fusion. In this context *known* implies recognized professional status, usually via a Ph.D., a publication record, and an institutional base. In contrast, someone with no scientific credentials, or none in the subject under discussion, will usually get a very skeptical reception. This was the case with Velikovsky, a psychoanalyst who ventured into celestial mechanics, ancient history, and astronomy, displaying a woefully uneven understanding of all areas. His innovations required intermittent disregard for the reliability of Newton's laws of motion and only episodic acceptance of the well-tested theories of gravitation and electromagnetism, with the invocation of speculative and totally qualitative suggestions for the operation of some new force. To just about every physicist and astronomer, Velikovsky's central ideas were patent rubbish, and professional reviews were vehemently critical. Yet Velikovsky drew continued support from humanists, social scientists, and some philosophers.

The cold fusion saga provides the clearest demonstration of the difference in reactions to revolutionary science that comes from inside or outside the scientific community. Within a week of the cold fusion announcement, and despite considerable skepticism, hundreds of scientists had dropped other tasks and were in their laboratories, devising new experiments or trying to replicate the Utah system. Others were busy reexamining nuclear theory. There was no similar reaction to Velikovsky. A small number of people made simple calculations to illustrate the extent of his errors, and a few of these refutations were published, but that was the extent of the reaction (apart from the responses of a few inveterate letter writers). There was, however, a *nonscientific* reaction that was notable, and I describe this when I review the whole episode in more detail in the next chapter.

Beneath the reactions of the varied participants and spectators are some fundamental questions and working methods and sets of assumptions. How can one tell the difference between science and pseudoscience, between genius and crank, between a great idea and a foolish conjecture? In some cases the confusion between science and its imitators is exploited, as with Krebiozen, where the "discovery" came from a doctor of uncertain provenance who had the strong endorsement of a major figure in American academic medicine (but not a cancer expert). Sometimes the distinction is clear (at least to the expert), accepted by all who have the same set of assumptions.

In other cases the acceptance comes almost as rapidly, as happened with the confirmation of predictions from Albert Einstein's General Theory of Relativity through observations during the 1919 solar eclipse. But in the case of Alfred Wegener's theory of continental drift, acceptance was long delayed because conclusive evidence was not at hand. The weight of professional opinion was strongly negative for decades. Wegener put forward his idea in 1912, but it took more than fifty years before accumulating evidence led to a professional flip-flop and plate tectonics (as it is now termed) became an accepted feature of geophysics. (Geophysicists in the Soviet Union, however, continued to resist this idea for another ten years.) The particular example of Wegener has led to a view among some critics of science that can be paraphrased something like this: "If the experts can be wrong (as with continental drift), and after many years a novel idea has finally been shown to be correct, then *this* current idea (erroneous though you may think it is) *is* correct, and it is only a matter of time before you come around to accepting it."

The inclusion of theories and their foundation of experimental results into the body of accepted science does not necessarily follow the route of popular impression. Philosophers and sociologists of science have explored this territory in great detail, and I draw on some of their writings. But I must add a word of caution: Most scientists remain blissfully unaware of that literature (or else disregard it). My views, as they emerge in this book, should be seen as representative of the views of typical working scientists, with whatever shortcomings this implies. So it will be necessary to review the workings of mainstream science—its methods, its own system of checks and balances, the criteria adopted by mainstream scientists—for these are the standards by which these researchers judge fringe science. Only then will we be able to understand the ways in which scientists view the subclassifications within the gray zone that surrounds the unquestionably genuine science. The analyses of many philosophers and sociologists of science may or may not be correct, but they play essentially no role in everyday science.

My current interest in fringe science was stimulated in 1967, when Immanuel Velikovsky was invited to Washington University to deliver the annual Sigma Xi

Society address at our honors day. I am not sure now why I did not go to hear his talk, but the critical reaction of one of my colleagues prompted me to look into the subject. Seven years later and through the invitation of a friend in the philosophy department, I found myself alongside Velikovsky in a panel discussion at a biennial meeting of the Philosophy of Science Association at the University of Notre Dame, debating his theory and its reception. It was a fascinating experience, one I describe later.

Velikovsky and his theory represent far more than simply one pseudoscientific episode among many; they illustrate typical aspects as well as some that are unique. The Velikovsky case has become a crusade, a continuing saga with a vast literature, much of it as nutty as the original thesis. He still has passionate defenders; there is still the occasional unanticipated eruption or aftershock as some innocent person strays into the minefield of writings and unexploded claims to draw fire from defenders of the faith and from those who are better informed. We do not have the space to explore all of the Byzantine intricacies of this fascinating case. If you wish to pursue it, I strongly recommend Henry Bauer's *Beyond Velikovsky,* which provides a good survey and an exhaustive bibliography. In the next chapter I focus on the essentials but also allow myself to indulge in a few digressions.

These cases and others like them may seem like little-known border skirmishes that do nothing to detract from science or deflect it from its current directions. Why then the fuss? One answer is that as long as the general nonscientific public cannot distinguish between science and pseudoscience, the possibility will exist for far more serious confusion, as will the possibility for consequent meddling and misuse of science. There can be great commercial and political pressure for the legalization of fake cancer cures like Krebiozen and the more recent Laetrile. After all, why should terminally ill and desperate people not be able to take any medication they wish? A ban on medications of unproven efficacy raises genuine ethical issues: Who has the right to make such decisions? But should any and every medication be available without restriction, without clearly demonstrated effectiveness? In the following chapters I examine many of these cases in more detail, with my main focus the evaluation of their scientific content and the ways in which these "discoveries" have been debated. But to avoid being taken too far afield, I do not discuss important aspects such as the medical ethics of cancer "cures." I do, though, ask why creation science (a religious theory of the origin of the world) should not be required as a part of the school science curriculum. And if so many people consult astrologers that many newspapers have daily columns of horoscopes, why should not national policy be similarly guided? In the final chapters I comment on the related matters of general scientific literacy and public policy and also suggest the application of a healthy dose of skepticism when we are confronted with extraordinary claims.

Notes

1. *Wall Street Journal,* 24 March 1989, 1.
2. *Reader's Digest,* March 1950, 139.
3. *Harper's,* January 1950, 19.
4. *Reader's Digest,* March 1950, 139.
5. *Bulletin for the History of Medicine* 58 (1984): 28.
6. *Nature* 171 (1953): 737.

2

Oldies but Goodies

Pseudoscientific products will always be with us. By definition these carry the illusion of science and sometimes have the endorsements of famous people (though not necessarily those with relevant credentials). In this chapter, I expand on two old examples of pseudoscience that I have mentioned rather briefly, Immanuel Velikovsky and the AD-X2 battery additive. Neither has genuine scientific content, and both provide clear examples of pseudoscience—one mostly academic in its claimed impact, the other because of its commercial pretensions and political support.

Worlds in Collision

Worlds in Collision appeared in April 1950, published by the respected Macmillan Company. It was an immediate success. First-year sales have been estimated at 75,000 copies. Even after more than forty years, a paperback edition is still in print. Reviews were extensive but sharply divided between high praise and sarcastic criticism. Why should *Worlds in Collision* have attracted such strongly opposed reactions? In part, this was the result of Velikovsky's ideas and his style of presentation. To a far greater extent, however, the opposition was ignited by the extravagance of the laudatory comments coupled with coordinated and aggressive marketing. What *was* it that Velikovsky was proposing that prompted such vigorous responses? In his preface Velikovsky described the starting point for his work:

> It was in the spring of 1940 that I came upon the idea that in the days of Exodus, as evident from many passages from the Scriptures, there occurred a great physical catastrophe, and that such an event could serve in determining the time of the Exodus from Egypt... . The historical-cosmological story of this book is based on the evidence of historical texts of many peoples around the globe.[1]

As a psychoanalyst who had studied with Wilhelm Steckel in Vienna, Velikovsky became interested in the Oedipus legend, and he saw in it similarities to the Egyptian pharaoh Akhnaton of the fourteenth century B.C. All of this led him to a comparison of the legends, epics, and written records of the Sumerian, Hindu, Chinese, Mayan, Aztec, Icelandic, Egyptian, and Hebrew peoples. He began to draw parallels and came to the conclusion that when these disparate and widely separated peoples referred to some happenings and memorialized them, they were probably describing real occurrences—and the *same ones*. Thus there had indeed been one worldwide flood (which Velikovsky associated with Noah) rather than a number of intense but local inundations that had also probably occurred. Similarly, when events in the heavens were mentioned, the descriptions should be taken literally as records of actual happenings rather than as allegories. So when Joshua commanded the sun to stand still, it was part of a sequence of real events with the stopping and resumption of the earth's rotation (which made the sun *appear* to stand still).

Having satisfied himself that the legends were describing ancient events seen simultaneously all over the earth, Velikovsky realized that the hitherto accepted chronologies of, for example, the Egyptian and Babylonian dynasties, needed adjustment in order to be brought into concordance. Because the legends of some peoples did not include all of the spectacular events noted by others, Velikovsky introduced the idea of "collective amnesia" to describe events so powerfully horrendous that their memory was "forgotten or displaced into the subconscious"[2] by a whole people. (But he did not explain why some societies remembered while others were unable to do so.)

What could have had so great an impact that it was recorded almost worldwide by some peoples yet was suppressed beyond recall by others? What were the events that were so spectacular that they found a permanent place in many tribal memories, to be handed down over the generations? Dorothy Vitaliano, a geologist at Indiana University, has shown how many of these tribal memories can be plausibly traced to unrelated and local geological features or events.[3] An examination of the topography of a region can show why it is subject to local flooding without recourse to a theory of global deluge. Velikovsky came to believe instead that remarkable celestial events had indeed taken place and been universally observed, and he finally produced the following scenario:

According to Velikovsky, the planet that we now call Venus did not exist in very ancient times. It emerged from Jupiter by some (unspecified) process and traveled through the solar system as a comet. Around the time of the biblical Exodus (fourteenth to fifteenth century B.C.), it approached so close that the earth passed through the comet's tail. The closeness of the comet caused the waters to part and allowed the Israelites to pass safely out of Egypt. Later, material from the comet tail was deposited in the atmosphere and fell to the earth as manna to sustain the Israelites in the wilderness. (The evidence that Velikovsky cited was the known presence of hydrocarbon molecules in comet tails, which he confused with carbohydrates that could be eaten.) Other cometary debris accumulated on the earth and makes up the petroleum reservoirs that we now tap and refine for so many uses. The Venus-comet had further encounters with the earth at fifty-two-year intervals. It also came near enough to disturb the orbital motion of Mars sufficiently that Mars came close to the earth in the seventh and eighth centuries B.C. A close approach by a comet caused the earth to stop rotating or its axis to tilt, making it seem the sun had stopped, an occurrence attributed to Joshua. During their close approaches the Venus-comet deflected Mars into its present planetary orbit. Velikovsky understood that gravity alone could not produce all of the effects that this history required, and he invoked electromagnetic forces.

In support of this theory, Velikovsky pointed to evidence other than that of the legends, and he also predicted what should be found if certain measurements were to be made. (Later, Velikovsky and his supporters claimed that these predictions had been confirmed and that these results provided the vindication that Velikovsky sought. These claims, however, have been contested and became the focus of further disagreements.) He believed that volcanic activity should have been triggered on the moon by its close encounters with Venus and Mars. Its surface should therefore show major signatures of activity (such as melting) less than 4,000 years ago. When their paths crossed, Venus and Mars exchanged atmospheres, so there should be hydrocarbons in the Martian atmosphere today. Because of its violent origin, Venus should still be hot, while Jupiter, site of the eruption, should emit radio waves.

Although *Worlds in Collision* was published in 1950, Velikovsky's ideas had been circulating through parts of the astronomical community for several years. In his 1946 pamphlet *Cosmos Without Gravitation*, Velikovsky had set out his novel ideas on gravitation and electromagnetism. He had written to Harlow Shapley, a leading astronomer at Harvard, hoping to persuade the people at a major observatory to undertake searches for what he was expecting. John J. O'Neill, science editor of the *New York Herald Tribune,* was much taken with Velikovsky's ideas and wrote about them as early as August 1946. One way or another, Velikovsky's ideas were known to many of the major astronomers by 1950. I make this point about prior knowledge because in a review the eminent Harvard astronomer Cecilia Payne-Gaposchkin rather foolishly stated that she had not read the book, and she was compared to the Roman prelates who refused to look through Galileo's telescope but did not refrain from condemning him and his ideas.

In *Harper's* of January 1950, before the official release date, Eric Larrabee took special note of Velikovsky's wide compass: "Dr. Velikovsky's work crosses so many of the jurisdictional boundaries of learning that few experts could check it against their competence."[4] (Some of us might have considered this as a flashing red light alerting us to problems ahead.) Larrabee did note that "Velikovsky willingly concedes that the behavior of the Earth and comet is not in accordance with the celestial mechanics of Newton," and he quoted O'Neill's opinion that Velikovsky's work "will stand as a challenge to scientists to frame a realistic picture of the cosmos."[5]

Larrabee also quoted Horace M. Kallen, former dean of the New School for Social Research, who had read the manuscript for Macmillan: He "would treat
[Velikovsky's work] as an extraordinary achievement of the scientific and historical *imagination*."[6] Fulton Oursler's condensation for the book section of the
March 1950 *Reader's Digest* emphasized the ways in which Velikovsky's recon-
structed chronology validated the biblical description of so many events, such as
Exodus, the flood, and the sun's standing still for Joshua.[7]

At the opposite end of the receiving line was Otto Neugebauer, dean of studies
of Babylonian mathematics and astronomy, who gave several examples of gross
inaccuracies in *Worlds in Collision*." He pointed out an excerpt Velikovsky had
translated from a German source then gave the original, with the comment, "Is
the author's knowledge of German so bad that he had to stick in a whole new sentence in a 'quotation'?" Neugebauer summed up *Worlds in Collision* as "an endless
repetition of errors, misquotations, irrelevancies and wild fantasies. The
Macmillan Company can congratulate itself on having found a very effective
method for extracting money from a wide public which will not be able to check
the factual basis of these 'works in confusion.'"[8]

In one of several critiques in the *American Journal of Science*, K. S. Latourette,
professor of oriental history at Yale, referred to Velikovsky's use of translations
from the Chinese: "He makes frequent and in some places extended mention of
the Emperor Yao. ... most [experts in Chinese history] would ... doubt whether
he ever lived."[9] A few pages further on, Chester Longwell, professor of geology at
Yale, wondered whether Velikovsky was showing "complete ignorance or [a] cava-
lier disregard of the evidence" and then went on to draw attention to the absurd
nature of one of Velikovsky's geological ideas.[10]

A full treatment of the errors in Velikovsky's analysis would require a book to
itself, and Bauer has done this. It must suffice here briefly to indicate the major
objections, many of which were raised in the early reviews and others derived
from space-age exploration. The orbital meanderings of Venus and Mars cannot
be explained in terms of our well-established understanding of gravity and the
principles of conservation of energy and momentum. Velikovsky gave partial recognition to this by suggesting, in a totally qualitative way, that electromagnetic
forces could be responsible. Our knowledge of the relative strengths of gravity and
electromagnetism on an astronomical scale simply do not support his conjecture.
There is no way in which today's Venus resembles the comets that we know. Comets are well described as dirty snowballs, loose aggregates that evaporate when
warmed as they approach the sun and so produce tails. Venus is a rocky and solid
body, its density similar to that of the earth, Mercury, and the moon. Exploration
of the lunar surface during the Apollo missions of 1969–1972 has definitively
shown that the surface has not been molten in historic times. Using the standard
radioactive method, rocks have been dated to 4.5 billion years, very close to the
ages based on examination of meteorites. Had those rocks melted during the in-
tervening millennia, radioactive gases could not have remained trapped, and the

derived ages would be off and would also show no consistency. Measurement of energy trapped in small crystals found in samples extracted from the top two meters of lunar soil has shown how that soil is gently "gardened" but not melted by the impact of meteorites.

Beyond these easily demonstrated flaws in the physics and astronomy of Velikovsky's scenario and theory, there are further shortcomings in his use of myths and legends, his translations, his application of outdated materials and innumerable selective quotations, his misunderstanding of Babylonian astronomy, and his idea of selective amnesia.

The confrontation between the reviewers from the sciences on the one hand and from the humanities and social sciences on the other was clear. Even as early as 1951, Velikovsky commented on the flavor of the dispute: "Are the humanistic and scientific approaches different? Scientists can calculate the torsion of a skyscraper at the wingbeat of a bird. ... They move in academic garb and sing logarithms.... We poor humanists cannot even think clearly.... We never take a step without stumbling; they move solemnly, ever unerringly, never a step back, and carry bell, book, and candle."[11]

What was notable in all the reviews and the resulting letters was the outraged unanimity of the scientific reviewers, each in his or her own field. Each excoriated Velikovsky's scholarship. Velikovsky's support came almost entirely from people not only without formal and relevant credentials, but more importantly (as shown by their own remarks) without an adequate understanding of the subject matter, which seemed in no way to inhibit their self-confidence or enthusiasm.

What at once distinguishes Velikovsky's writings from almost all other pseudo-scientific treatises is the appearance of scholarship—most pages have several footnotes, attesting to his impressive use of library resources. It is clear that this helped to persuade his nonexpert supporters. What we have seen, though, as evidenced by the many expert reviews, is that Velikovsky misused these sources. I could fill a very large volume with detailed examples of Velikovsky's lack of understanding, his use of innuendo, his errors and misquotations and selective uses of quoted sources in his books and later articles.[12] Let me mention here only a very few that I found in my own explorations.[13]

Velikovsky quoted R. A. Lyttleton, an eminent astrophysical theorist, as having written that "Venus must have erupted ... from Jupiter."[14] When we go back to Lyttleton, we fail to locate the word *must*. Lyttleton explores *possible* scenarios, using heavily qualified phrases: "If ... then ... could."[15] A second example: Velikovsky claimed that he had predicted the retrograde rotation of Venus.[16] (Most rotation in the solar system is counterclockwise if one imagines viewing from a position far above the North Pole. "Retrograde" rotation refers to rotation in the clockwise sense.) Velikovsky's actual prediction was that the rotation of Venus "might well be retrograde." This is as useful as predicting that the outcome of a coin flip might well be heads. A last example: In his 1955 book *Earth in Upheaval,* Velikovsky graciously acknowledged Albert Einstein:

The late Dr. Albert Einstein ... gave me much of his time. He read several of my manuscripts and supplied them with marginal notes. ... In the last weeks of his life he reread *Worlds in Collision* and read also three files of "memoirs" on that book and its reception, and expressed his thoughts in writing. We started at opposite points; the area of disagreement ... grew ever smaller. ... At his death ... his stand then demonstrated the evolution of his opinion in the space of eighteen months.[17]

In contrast to this view, we have the remarks of I. Bernard Cohen, the Harvard historian of science who visited Einstein only two weeks before his death and described their conversation:

The subject of controversies over scientific work led Einstein to take up the subject of unorthodox ideas. He mentioned a fairly recent and controversial book, of which he had found the nonscientific part ... interesting. "You know," he said to me "it is not a bad book. No, it really isn't a bad book. The only trouble with it is, it is crazy." I replied that the historian often encountered this problem: Can a scientist's contemporaries tell whether he is a crank or a genius when the only evident fact is his unorthodoxy? "There is no objective test," replied Einstein.[18]

And here we have the essential point of so much of what we have yet to treat. What are we to make of all of this? How far does one go in retaining an open mind to new ideas, especially when they are supported by such substandard scholarship? If nothing further had happened, Velikovsky would probably be only one of the many pseudoscientists who have turned up over the years. Instead, the appearance of *Worlds in Collision* led to a reaction unique in the history of modern pseudoscience.

Not content with ridiculing Velikovsky and his book, a number of astronomers took their protests directly to the publishers. They threatened to boycott Macmillan textbooks unless the company dropped publication of *Worlds in Collision*. Pressure came from Harlow Shapley and Dean McLaughlin of the University of Michigan. The return of textbooks and the refusal of some professors to review new manuscripts also shaped Macmillan's response: Within two months of publication, and with the book already on the best-seller lists, Macmillan, with the reluctant acquiescence of Velikovsky, transferred publication to Doubleday, which had no textbook division and was thus beyond this sort of pressure. In the wake of this episode, James Putnam, an editor at Macmillan, was fired, as was Gordon Atwater, director of the Hayden Planetarium at the American Museum of Natural History in New York, who had been planning a special presentation highlighting Velikovsky's ideas.

The resulting self-inflicted damage the scientific community has sustained is immense in the eyes of its critics. It is this intolerant attempt at censorship that has kept the embers of the Velikovsky case aglow, repeatedly attracting the angry comments of humanists and social scientists who have been critical of science. After the *Mariner 2* space probe had visited Venus in 1962, Velikovsky's supporters claimed that some of his predictions had been verified. Such claims do not themselves survive scrutiny, but this produced further allegations of a concerted censorship to deny Velikovsky his due credit.

There was an outburst of interest a few years later with extensive treatment in the *American Behavioral Scientist* and, after a few more years, in a complete issue of the *Yale Scientific Magazine,* punctuated by some acid exchanges in the pages of the *Bulletin of the Atomic Scientists.*[19] The 1970s saw yet another resurgence of interest in Velikovsky's ideas. *Pensée,* a new magazine, was devoted to Velikovskian studies; ten issues appeared, with a circulation of somewhat over 10,000. When it ceased publication, it was replaced by *Kronos,* which continued to appear until 1988. During this surge of interest in the 1960s and 1970s, Velikovsky was invited to speak at a number of colleges, and some even offered courses centered on his work.

In 1974, with a record of over twenty years of hostility, the scientific community gave Velikovsky a platform at the annual meeting of the American Association for the Advancement of Science. The papers presented there, with the notable exception of Velikovsky's, appear in the book *Scientists Confront Velikovsky.* (Velikovsky refused to provide a manuscript because of disagreement over how to handle rebuttals.) In the same year the Philosophy of Science Association (PSA) arranged a panel discussion for their biennial meeting. In addition to Velikovsky and myself, the panelists were two strongly pro-Velikovsky philosophers of science whose contributions were singularly irrelevant.

Of course, neither meeting could satisfy Velikovsky, convince him of his mistakes, or remove his deep sense of hurt through such long rejection. His attitude, both toward his own work and toward scientists, was firmly set. When we were introduced just before our panel assembled at the PSA meeting, he asked me, "And which of my books have you read, if any?" During my presentation, I described inaccuracies in his writings. He challenged the accuracy of my quotations ("Where did I write that?") then switched to another point when I produced a photocopy of the pages in question. I moved to another example of misuse of the literature, and we went through this routine again. It was impossible to get a substantive discussion on any scientific point. His contribution to the panel discussion consisted of a long diatribe that his work had not been taken seriously and a complaint (somewhat justified) about the treatment he had received from the scientific community. After our formal presentations, a member of the audience asked him what he would change in his writings now, after more than twenty-five years of space exploration and discovery. His reply: "Not one word."

The papers presented at the PSA meeting were published as well, but again Velikovsky did not submit a manuscript. He did, though, publish numerous articles in *Pensée* and *Kronos* covering the same ground. As new results came in from the space sciences, Velikovsky and his supporters seized upon those that they thought demonstrated the correctness of his theory and predictions. Yet when examining Velikovsky's interpretations of these "validations," I have again found a trail of misquotations and misunderstandings. Nevertheless, claims that he has been shown to be correct still surface. It is clear that to some hard-line supporters, no demonstration of error will be acceptable.

We are still left with the questions: Why was the boycott threatened? And why have Velikovsky's ideas attracted such tenacious support? As best I can determine, there was a spontaneous expression of outrage from many scientists that so distinguished a publishing house should have produced such a patently pseudoscientific work whose promotion traded on the public's general inability to judge critically, especially when the book was promoted by a barrage of glowing (though worthless) reviews. Confusion over the correct trade classification of the book continued for years, starting with Macmillan's inclusion of the book in the science section of one catalog; another publisher categorized it simply as nonfiction.

And why the prolonged support? A combination of reasons is probably the answer. The apparent validation of the literal truth of the accounts in the Bible had great immediate appeal, especially since Velikovsky supported his text with an impressive use of citations. It is also clear to me, from personal conversations and many articles, that there was a deeper reason for the backing of nonscientists: a deeply rooted hostility to science that draws its strength from both ignorance and misunderstanding. Some people resent the successes of science and may even blame it for any number of ills in the world. The Velikovsky case and especially the boycott simply provided one more opportunity to attack science. In recent years some Velikovsky backers have claimed support for his theories in the suggestion that a catastrophic event such as a cometary impact might have been associated with the demise of the dinosaurs. It seems as if they believe support for almost any catastrophe theory lends credence to Velikovsky's theory.

The consequences of the events of 1950 are clear. Though the Velikovsky case is now receding, I would not be surprised to find the issue of the boycott being revived again, to serve once more as the weapon with which to attack the scientific community for its attempts to suppress the truth when it comes from someone not a member of the club.

The AD-X2 Battery Additive Case

A central aspect of the Velikovsky case was the rejection of informed opinion by a large number of nonscientists, and we find this same phenomenon in our next case study. This is the strange (but comparatively short-lived) episode known as the AD-X2 battery additive case, in which political intrusion ensured that the early outcome would not depend only upon the scientific merits.[20]

In the years immediately following World War II, a shortage of lead affected many industries, especially those that manufactured automobile batteries. In response to this commercial opportunity, over 100 proprietary products designed to improve the performance and/or extend the life of auto batteries were marketed.

They carried names such as Duble Power, Bat-Re-Nu, and Nu-Zip. Tests carried out by the National Bureau of Standards (NBS) as far back as 1931 had shown that no battery additive of this type produced any beneficial results, and some were actually harmful. While the secret formulas varied, the major ingredient was usually Epsom salts. The National Better Business Bureau (NBBB) and consumer organizations publicized these findings.

The promoter of one of these proclaimed elixirs, Jesse M. Ritchie of Oakland, California, was described as "a self-educated engineer, a certified bulldozer operator" who was listed as "'Psychologist-Specialist in Alcoholism' in the phone directory" and claimed to have a doctorate in psychology from the College of Universal Truth, Chicago.[21] In 1947 he became a partner in a company that manufactured battery additives, and he engaged Dr. Merle Randall, retired professor of physical chemistry at the University of California, to help with the development of an improved formulation. He named the new product AD-X2. Rather than go through the patent application process, with its required disclosure, Ritchie kept the AD-X2 composition a secret.

Given Ritchie's vigorous promotion of AD-X2 and the history of similar products, it was no surprise that consumer organizations became involved and the controversy continued until well into the 1950s. For example, in a report on battery improvers, the *Consumers' Research Bulletin* commented on "promoters seeking to make a quick dollar by a pseudo-scientific 'invention.'"[22] In 1948, to substantiate his claims that *his* product was effective, Ritchie requested that the NBS test AD-X2; the NBBB also asked for testing, though obviously for different reasons. The NBS, however, would not undertake commercial testing at the request of an individual manufacturer or vendor, nor would it identify by name the products that it tested at the request of a government agency. By June 1951 the NBBB filed a complaint with the Federal Trade Commission (FTC) and found itself being accused by Ritchie of interfering with his business. The FTC received a second complaint a year later, this time from the Association of American Battery Manufacturers. Meanwhile, Ritchie had solicited support for NBS testing from the Oakland Chamber of Commerce and from Senator William Knowland of California. In March 1950 the FTC requested technical advice from the NBS, despite the bureau's policy, and a month later the NBS reported that no "significant reduction in harmful sulfation" had been found, in other words, AD-X2 was worthless. For legal reasons, the FTC was not satisfied with this NBS report and requested a second evaluation from NBS. The results, produced in July 1952, reaffirmed the findings of the earlier tests.

During all of this, the Post Office Department had also requested evaluation. Armed with the NBS findings, the Post Office had notified Ritchie in March 1952 that he was to appear at a hearing in Washington on charges of "conducting an unlawful enterprise through the mails."[23] Ritchie then appealed for help from the Senate and House Committees on Small Business. The staff director of the House committee asked the NBS to undertake a new test, this time following a protocol set out by Ritchie. The NBS carried out the test in June 1952; again the results showed AD-X2 to be worthless, and Ritchie criticized the test procedures.

In October a staff member of the Select Committee wrote to Dr. Julius Stratton, provost of MIT, requesting additional tests. MIT carried out its tests by November 1952, and its report was quoted as listing eight effects of AD-X2 that were found to support the claims of the manufacturer. Dr. Allan Astin, director of the NBS, pointed out that the MIT tests had used AD-X2 in such dilution "that it appears to be of no significance whatever in normal storage battery operation."

Then, in February 1953, the decision was handed down from the U.S. Post Office that Ritchie was to stop using the mails. One week later the postmaster general suspended that decision (and in August canceled it). Further political intervention came with the arrival of a new administration in January 1953, when Secretary of Commerce Sinclair Weeks agreed both with the suspension of the Post Office decision and also with the requested resignation of Astin from his post at the NBS. Weeks also suspended all NBS circulars dealing with battery additives and stated that he would request a new and completely independent test by the bureau. Astin's firing brought about the intervention of Dr. Detlev Bronk, president of the National Academy of Sciences, who urged Weeks to postpone Astin's dismissal. In May Weeks asked the academy to undertake the testing, and in October the academy's Committee on Battery Additives released its findings.[24] Its major conclusions were that the NBS had carried out excellent testing and that AD-X2 was "without merit." The academy report was somewhat critical of the MIT testing but was much more critical of the publicized interpretation of that report by a Select Committee staffer: The academy said the MIT tests did not support AD-X2. At the same time an internal review within NBS, carried out at Weeks's direction, to study the role of NBS in similar situations, recommended that NBS not be involved in future commercial testing. By August Weeks had changed his mind and decided to retain Astin as head of the NBS.

One might reasonably be excused for thinking that the issue would have been decided by this time, but a science that had been insulted continued to be dragged around the arena of political maneuvering. The Senate Select Committee pursued its hearings "to help create an atmosphere ... that is conducive to the welfare of small-business enterprises."[25] Letter-writing campaigns were directed at members of Congress. The Justice Department was brought into the fray to prepare an antitrust case against the Association of American Battery Manufacturers, charging it with conspiring to prevent the resale of lead from used batteries.

The case eventually faded away. The Post Office set aside its order, and the FTC similarly withdrew its order. Committee hearings had been inconclusive. Claiming interference with the conduct of his business, Ritchie unsuccessfully sued the Post Office.

Today, after more than forty years, what are the issues and lessons we can see? First, of course, we should look at the scientific merit of the claimed "invention." There was none. Second, we should examine the method of promotion: By avoiding the necessary disclosures of the patent process, Ritchie was initially able to evade any demonstration of the effectiveness of his product. When consumer protection pressures led to testing by the NBS, attempts were made both to discredit the tests and to bypass them. Third, the government received many testimonials in support of AD-X2, and these carried sufficient weight, in the eyes of some members of Congress and the administration, to more than balance the expert testimony of the NBS. As had been the case with Krebiozen, a few scientists did give their support to the product. Besides Randall, Ritchie's early consultant, Dr. Keith Laidler, associate professor of chemistry at Catholic University, a technical adviser to the Senate Select Committee, and a subsequent consultant to Ritchie, was highly critical of the NBS tests.

Probably the most distressing aspect of this case was that it attracted political involvement even though evidence indicated the worthlessness of AD-X2. The Select Committee seemed more concerned with protection of a small-business constituent. Its motivation was well summarized in the staff report in *Technical Information for Congress*. This document clearly set out the very different ways in which disagreements are resolved in science and in political life, and the ways in which different audiences weigh expert testimony against testimonials from users. The report summary is breathtaking in its assessment of the role of science:

> Political decisions and policies are sometimes found necessary to mediate, postpone or circumvent the effects of harsh and arbitrary findings of science that impose unacceptable obligations or conditions on the electorate or the individual. ... Out of the case came the decision that the primary role of science in commerce should not be to regulate the quality of products to protect the consumer but to discover the truths of nature, and use them more particularly to create additional products or human satisfaction and entrepreneurial exploitation.[26]

Before we relax and think that all of this happened forty years ago and that conditions are different now, we should pause to note that many of the same issues and pressures run as a central theme through a long string of other issues, from radioactive fallout from weapons testing to today's environmental issues, such as air quality and the effects of electromagnetic fields. Behind all of these confrontations is a deep-seated suspicion of expert advice and a willingness to accept lay testimonials as equally valid, as well as a sweeping public lack of understanding of the methods and limitations of science and the quality of its conclusions. Legislative bodies have too frequently been willing to intrude their judgments even when the scientific advice has been loud and clear. We will, you can be sure, encounter these suspicions again.

Notes

1. *Worlds in Collision* (New York: Dell, 1965), xvi.
2. Ibid., 330.
3. *Legends of the Earth* (Bloomington: Indiana University Press, 1973).

4. *Harper's,* January 1950, 19.

5. Ibid., 23, 24.

6. *Harper's,* January 1950, 20 (my emphasis).

7. *Reader's Digest,* March 1950, 139.

8. *Isis* 41 (1950): 245, 246.

9. *American Journal of Science* 248 (1950): 584.

10. Ibid., 588.

11. *Harper's,* June 1951, 66.

12. C. Leroy Ellenberger and Robert Forest, among others, have drawn attention to this weakness of Velikovsky. See H. H. Bauer, *Beyond Velikovsky* (Urbana: University of Illinois Press, 1984), for references.

13. *PSA 1974: Proceedings of the 1974 Biennial Meeting of the Philosophy of Science Association* (Boston: D. Reidel, 1976), 477.

14. *Yale Scientific Magazine* 41 (1967): 8.

15. *Man's View of the Universe* (Boston: Little, Brown, 1961), 36.

16. *Yale Scientific Magazine* 41 (1967): 9.

17. *Earth in Upheaval* (New York: Dell, 1955), vii.

18. *Scientific American,* April 1955, 69. But see also *Scientific American,* September 1955, for a letter from Dr. Otto Nathan, executor of the Einsten papers, and also Velikovsky's *Stargazers and Gravediggers* (New York: Morrow, 1983), 296, taking issue with Cohen.

19. *American Behavioral Scientist* 7, September 1963; *Bulletin of the Atomic Scientists,* April 1964, 38.

20. Unless otherwise noted, all quotations for this case come from U.S. House, *Technical Information for Congress: Report on the Subcommittee on Science, Research and Development of the Committee on Science and Astronautics,* 92d Cong., 1st sess., 1969, 4. Rept. 57–307, 14–60.

21. Ibid., 19.

22. February 1951, 17.

23. *Technical Information,* 28.

24. *Science,* 117 (1953): 3 and 409, and 118 (1953): 683.

25. *Technical Information,* 31.

26. Ibid., 18.

3

Hesitant Revolutions—
Successful and Failed

I chose the examples in the preceding chapter because they represented an extreme version of fringe science. Neither Velikovsky nor Ritchie had acceptable credentials in the fields in which they claimed to have made discoveries. Neither claim attracted the serious attention of any scientists; indeed, the role of the scientists was largely confined to debunking from the start, through reviews (in the case of Velikovsky) and only reluctantly (as with the scientists at NBS and MIT, when called upon to test AD-X2). Both cases drew vehement support from many people who clearly did not possess the necessary expertise.

In contrast the two cases I present in this chapter did engage the professional attention of scientists. Curiously, both examples, continental drift and cold fusion, started with novel ideas from scientists working outside their established fields. Initially, both commanded significant attention. Beyond that, their histories diverge. After attracting considerable interest, the theory of continental drift languished for many years until the application of new experimental techniques yielded results that rapidly tipped the balance. The claims for the production of fusion at room temperatures, though, have not been sustained, the surge of apprehensive excitement has long since dissipated, and cold fusion seems destined to be remembered only in books like this.

Continental Drift

In popular imagery major discoveries are announced by a shout of "Eureka!" as the inspirational flash shoots through the inventor's brain. Archimedes leaped from his bath; Newton was struck by both apple and insight; and Thomas Edison emerged from his attic or basement or wherever, carrying the light bulb, still the cartoon insignia of sudden inspiration. But, so the myth continues, the independent genius is often ignored or even ridiculed before the belated general acceptance arrives. This delay in recognition has sometimes been used to defend the proposals of pseudoscientists or inventors with the claim that experts have been proven wrong before and that the current proposal, whatever it is, will also, in time, be shown to be correct, to the discomfort of the experts who know this but do not wish to relinquish their current orthodoxy and exalted status.

This was a theme in the Velikovsky debates and an undercurrent in the AD-X2 case, where expert opinions were repeatedly challenged. At least superficially, the reception given to the idea of continental drift would seem to provide strong support for this thesis. Closer examination reveals a very different history.[1]

By now, it is almost universally accepted that the earth's major land masses were once part of a giant continent, Pangaea, that had the northern and southern parts of Laurasia and Gondwana. Some hundreds of millions of years ago, this supercontinent gradually broke into parts that drifted apart and continue to move, even today. The name now given to this geological process is *plate tectonics*, intended to describe a larger range of phenomena than the original term, *continental drift*.

Even the most superficial comparison of the eastern coastline of South America with the western outline of Africa suggests their original proximity. Converting this simple observation to an accepted scientific theory has taken many years and required more geophysical evidence than the simple jigsaw puzzle approach. As we will see, the tale of hesitant advance has many similarities to the Velikovsky case but with some revealing contrasts.

Because of the apparent solidity of the earth, people long considered it immovable, the center of the universe, its continents unchangeable. There was, of course, Plato's story of Atlantis, a lost continent, but this did not form part of any general theory of earthly evolution. Indeed, even the idea of the continents' having moved could not have arisen until there were reliable maps so that the similarity of the outlines could be seen. Africa was probably circumnavigated by the Phoenicians, but even if they drew up any charts after those voyages, none has survived. South America remained hidden from European awareness until the voyages of discovery in the fifteenth and sixteenth centuries. (It was during those voyages that the explorers delineated and named the constellations in the southern skies that cannot be seen even from southern Europe and the Middle East, from which region the classic Western constellations had been named.)

By 1620 Francis Bacon was drawing attention to the mirror image relationship of the Atlantic coastlines, but sustained examination had to wait for about 200 years. By then, geology was developing, and the basic idea of an immutable earth was being questioned. Scientists were putting forward theories to explain processes that could have changed the face of the earth. During the eighteenth century, the study of geological strata and their included fossils was leading scientists to challenge the accepted estimate of the earth's age, which had been based on the tally of generations as enumerated in the Old Testament. In his *Principles of Geol-*

ogy in 1830, Charles Lyell set out a view that held for many years: that the earth's surface had undergone changes through cycles of uplift and erosion.

As the end of the century approached and even with widespread acceptance of the notion that the earth had not always had its present form, there was still room for competing theories. The problem of the age and evolution of the earth had been sharpened as theory was being adapted from physics. Out of sophisticated calculations, two major schools of thought emerged. One group believed the continents had been formed far in the past, when cooling and contraction had resulted in the uneven surface, much as an apple wrinkles when drying. Since that remote time, the continents had been essentially unchanged. The other major opinion held that *because* of the cooling and contraction, geological processes in the past differed from those today. Accordingly, simple extrapolations back in time were unreliable.

All theories had to deal with the evidence that some fossils of the same species were found in South America, Africa, and even Antarctica. With the idea of evolution gaining ground, it seemed very unlikely that identical species could have developed independently in such widely separated locations. To meet this dilemma, it was proposed that today's continents had once been connected by "land bridges" across which migration could have taken place before the bridges subsided.

At the center of the many debates, scientific and otherwise, was an assumption that came to be known as uniformitarianism, the belief that physical and chemical laws and processes known today also operated in the same ways in the remote past. In the competition between science and religion, this assumption has been repeatedly challenged: Perhaps different processes had operated, in which case the history of the earth could be very different from current scientific ideas and perhaps even consistent with the biblical account. We find Velikovsky and his proponents using the identical argument. Two of my fellow panel members at the Notre Dame meeting were philosophers of science who refused to accept the idea of uniformitarianism.

Uniformitarianism is not an assumption to be adopted lightly. In modern science Newton introduced it in his 1687 *Principia*, at the start of Book 3, "Rules of Reasoning in Philosophy." "We are to admit no more causes of natural things than such as are both true and sufficient to explain their appearances. Therefore to the same natural effects we must, as far as possible, assign the same causes."[2] We use this assumption repeatedly in astronomy. When we examine the light from distant stars and the even more distant galaxies, we are also looking back into the past. The light arriving today was radiated millions or billions of years ago. We interpret conditions in those remote times in terms of laws that we have discovered, using the values of the physical constants as we have measured with great accuracy. We can devise tests using the "old" light to check whether our current scientific laws operated in the familiar ways and with the same physical constants much closer to the origin of our universe. So far the results have always confirmed the stability of those laws and the values of those constants.

There are other aspects of uniformitarianism. As used in geological debates, it has been taken to imply not only the operation of the same physical processes but also their operation *at the same rates* as now. This is a much narrower assumption. There is also always the possibility that new physical processes may be discovered so that calculations of past conditions will be proven wrong because they are incomplete. But that is something we live with at all times. We keep using our currently known laws yet remain alert to possible problems. Without some assumptions about the nature and stability of those laws, we would have no objective way of choosing between countless alternatives.

By 1900 the discovery of radioactivity added to the woes of contraction theory. Radioactivity provided a novel and accurate method of geological dating, suddenly expanding the known geological time scale to billions of years. Radioactivity was found to be a source of heat as well as penetrating radiation. Estimates of the total amount of the radioactive uranium and thorium in the earth and the heat they produce called into question the earlier calculations of the earth's cooling. There seemed to be too much radioactive heat energy released to allow for the contraction and cooling that had been assumed.

Because of the inconclusive nature of the theories then current, a substantial number of scientists thought that what geology really needed was much more data, not sweeping theories. It was at this juncture that Alfred Wegener presented his theory of continental drift.

Wegener had studied physics, then took his doctorate in astronomy; his thesis was on the Alfonsine Tables of planetary positions, which had been calculated in the thirteenth century and had played a central role in medieval astronomy. His subsequent interest was in meteorology, and in 1906 he was selected to accompany a Danish expedition to Greenland. He later held a faculty position at the University of Marburg and accompanied a second Danish expedition to Greenland in 1912.

Wegener's ideas seem to have been stimulated initially by the well-known comparison of the American and African Atlantic coastlines. Wegener introduced his theory at the 1912 meeting of the German Geological Association; his paper was published later that year and by 1915 was expanded into a book, *The Origin of Continents and Oceans*. Wegener's thesis went far beyond simple contour comparison. In the first place, he compared the edges of the continental shelves and found a better fit than with the conventional coastline matching. He also undertook a worldwide analysis of elevation levels of the earth's surface, both continental and submerged. As he pointed out, if the earth had cooled from an original smooth and molten sphere, one would expect to find the elevation levels clustering around some average value, with decreasing numbers far from this average. Instead, there turned out to be two dominant levels, those of the continents and those of the ocean floors. True, there were major mountain peaks and ocean abysses, but those were the exceptions. The contour levels certainly did not match what was expected from cooling.

Among the fossils were some of mesosaurus, a small reptile, which were found only in Brazil and South Africa, suggesting an early proximity of those regions. There were other fossils, notably those of *Glossopteris* ferns, which were found in South America, the Falkland Islands, South Africa, Australasia, and also the Antarctic, suggesting that they had at one time shared a common climate much warmer than that of the now frigid Antarctic. Wegener also drew attention to modern measurements of longitude that appeared to indicate a shift westward of Sabine Island (near Greenland) of nearly a kilometer between 1823 and 1907. Finally, Wegener took issue with the land bridge theory. If, as was generally thought, the less dense continents floated atop the denser mantle, the next lower layer of the earth's crust, then presumably the land bridges also consisted of light, continentlike material. Wegener pointed out that it was unreasonable to expect that those bridges had sunk when they had done their job. In sum, Wegener's theory was based on far more than simply the obvious continental shapes.

Initial reactions were mixed. Wegener's views became more widely known only after World War I, with the 1922 English translation of his book, though the Berlin Geological Society had held a symposium on drift in 1921. There were further conference discussions at the Geological Society of France, the Geological Section of the British Association for the Advancement of Science, and the Royal Geological Society in 1923. In a careful but critical review of drift in *Nature,* Philip Lake expressed disappointment that "Wegener has given free play to his imagination"; it was clear that he was not persuaded.[3]

An important contribution came from Harold Jeffreys in his 1924 book *The Earth.* Jeffreys's work was far from practical geology; he was a highly regarded applied mathematician who approached the drift hypothesis as a problem in mechanics, considering the strength of the earth's material, the forces that were known, and the resulting drifts that seemed possible. He calculated the forces needed to move the continents against the resistance of the underlying material and found, to his own satisfaction, that drift was impossible. He summed up his opinion in obvious terms: "The assumption that the Earth can be deformed indefinitely by small forces, provided only that they act long enough, is therefore a very dangerous one, and liable to lead to serious errors."[4]

For a long time Jeffreys's objections remained as influential obstacles to drift theory. He asked what seemed to be reasonable questions: Why did the continents drift? What force was responsible? In the absence of a clear identification of quantitative agents for change—which even Wegener admitted—it was hard to argue with the results of such impressive mathematics.

Another debate at the 1928 New York meeting of the American Association of Petroleum Geologists failed to move any of the alignments of the opposing sides. By this time, as Le Grand has well described, the shape of the debate was firmly set. In Europe there was a tradition of broad theorizing, cutting across the lines of subdisciplinary specialities; it was acceptable to base theories on careful syntheses from the primary literature. In contrast, in Britain and the United States, geology was progressing in increasingly separate areas of specialization, and greater weight was placed on the results of direct fieldwork. Theorizing was not highly regarded.

With many professionals undecided and many clearly opposed to the drift theory, Wegener had few outright supporters. Notable among the supporters was Alec L. du Toit, a South African geologist, and Arthur Holmes of Edinburgh. Du Toit had been persuaded by Wegener's use of the fossil evidence with which he was so well acquainted from his own fieldwork in South Africa and Australia. Holmes had responded to some of Jeffreys's criticisms and also introduced the important suggestion that drift was the result of convection currents, driven by heat from radioactivity.

Among the many comments, probably the most interesting (for us) came from Chester Longwell of Yale. In the 1928 symposium he reserved his judgment, noting that "many of us, however, are not yet prepared to accept literally Wegener's interesting analogy of the torn paper as applied to Atlantic coasts" but that "the only objective attitude is one of 'wait and see.' ... The true solution ... might well outrage some of our preconceptions."[5] He repeated his cautious stand in some articles in 1944, and we find a similarly critical approach in his review of *Worlds in Collision*.[6]

What we see in Longwell is a scientist who did not reject a truly radical theory out of hand but was willing to examine it critically, at the same time and quite correctly holding it to a high standard of demonstration. Over a long period Longwell kept his mind open. This approach was scientific in the true sense. We should recall his review of Velikovsky's *Worlds in Collision*, when his scientific sensibilities had been outraged. He was *not* unwilling to look at any and every new theory—only those that had a substantial base of initial plausibility.

Continental drift remained in its state of suspended credibility until after World War II. Intensive research as part of the war effort had produced many technical advances that became generally available to science. Electronics blossomed, its postwar application leading to the explosive growth of radio astronomy. Research in submarine detection produced echo-sounding devices that physicists later applied to geology. Patrick Blackett at Manchester University and later Imperial College in London, winner of the Nobel Prize in physics for his cosmic ray discoveries, was responsible for greatly improving the sensitivity of a device for measuring weak magnetic fields. With S. K. Runcorn of Cambridge University and the University of Newcastle-on-Tyne, he was able to measure the minute magnetism that was frozen into certain rocks as they cooled. The direction of the magnetization showed the direction of the earth's magnetic field at the time of cooling, and comparison of these directions in rocks from both sides of the Atlantic showed intercontinental similarities that were quickly interpreted as supporting the theory of drift. Parallel developments applying sonic methods to marine science greatly expanded seafloor exploration. Improved methods were brought to submarine seismology, measurements of heat flow and gravity variations.

The overall picture that emerged from this technological revolution showed an ocean floor that was not a remnant from long cooling and contraction but a very young floor continually being generated by material welling up at midocean ridges. Holmes's idea of convection currents was revived and refined. This process, plate tectonics, included the idea that the ridges were the sites of new material coming up from the underlying mantle, with older seafloor material being drawn down at ocean trenches, thus returning to the mantle. Riding atop this slowly moving seafloor were the continents.

One can trace the relatively rapid switch in the prevailing professional views through conferences in the late 1950s and early 1960s. During this period the International Geophysical Year (1957–1958) had led to coordinated worldwide surveys. Le Grand has described the results: "Beginning in 1966, sentiment swung rapidly to Drift. ... By the early 1970s, a new version of drift, plate tectonics, had become the established program. The 'revolution' was complete."[7] At this point we should take stock. What does this case study show us? It shows the conservatism of the scientific community, its unwillingness to rush to accept a radical theory for which there was some—though far short of conclusive—support. While most professionals were not persuaded, a few believers persisted in their support but, it must be said, with no new evidence to offer for many years. There certainly was not total neglect of the drift hypothesis, but the revolution (if any) came only with the introduction of totally new experimental techniques. Thereafter the change was swift. It is probably fair to say that without the postwar instrumental revolution, we might still be listening to reruns of the old arguments, waiting for the drift hypothesis to be established.

As a postscript we should note that well into the 1970s, the drift theory was resisted in the USSR.[8] The main opposition came from Vladimir Beloussov, director of the Department of Geodynamics in the Institute of Earth Science of the Academy of Sciences, who had developed his own theory. Although some Soviet scientists accepted the newly developed plate tectonics theory, Beloussov continued to oppose it even as late as 1989.[9] Even as there is general acceptance in the West, there are still some pockets of resistance today. However, the new dogma of plate tectonics seems now to be firmly in place.

Cold Fusion

After many years the frog of continental drift turned into the charming prince of plate tectonics. The reverse seems to be the case with the claims for the discovery of cold fusion, whose transformation has been rather more rapid. It seems improbable that the good fairy would arrive on a return visit and restore respectability if not beauty to what might be better described as the cold fusion turkey. This ongoing scientific soap opera carries with it some lessons or warnings that I examine later, along with a variety of other instructive cases. We can start with a short review of the physics of nuclear fusion.

Every atom consists of a nucleus made up of protons and neutrons, surrounded by a number of electrons held around the nucleus by an attractive electric force between the electrons and protons. Within the nucleus the protons and neutrons are bound by their nuclear force, which is enormously stronger than the electric force of repulsion between the like-charged protons. The simplest nucleus is that of hydrogen (^{1}H), with only a single proton. Deuterons (^{2}D), heavy isotopes of hydrogen, have one proton and one neutron in each nucleus. The next heavier element is helium, mostly of the ^{4}He variety, with each nucleus containing two protons and two neutrons; there is also a less abundant isotope ^{3}He with one neutron less. In nature, there are close to 100 different elements, many with several isotopes.

Most stars consist of hydrogen, usually with less than about 25 percent helium by mass and a small percentage of heavier atoms. Held together by gravitational attraction, the stars have pressures and temperatures at their centers that are so high that all of the atoms are ionized, that is, electrons are stripped away from their nuclei, and the star core consists of a plasma of electrically charged particles, nuclei and electrons. At the center of the sun, the temperature is about 15 million degrees Celsius, and the protons are all moving at high speeds. When they collide, they usually rebound, but a minute fraction collide so violently that their electric repulsion is overcome and they fuse to produce deuterons, releasing energy in the process. This is the first stage of the fusion process that keeps our sun and most of the stars glowing. Further collisions can then produce the helium isotopes, ^{3}He and ^{4}He. Every second, 600 million tons of hydrogen fuse in the sun to produce 10^{38} nuclei of ^{4}He. In addition to the great quantity of heat they generate, the fusion reactions produce neutrons and gamma rays.

The role of fusion in stellar interiors has been known since the 1930s. Explosive fusion has been achieved on earth in hydrogen bombs, but we would like to achieve a less violent fusion so that we can control and harness the energy released. Controlled fusion has long been the aim of international research programs because it holds the potential of a source of energy for electric power without the production of the large amount of undesirable radioactive waste that inevitably accompanies current nuclear reactors that work by the fission of uranium. In addition, the primary fuel for fusion, hydrogen and its isotopes, is in plentiful supply because of the vast reservoir in the waters of the oceans.

The basic physics is well understood, and the research programs have almost all concentrated on achieving and sustaining high temperatures as the means of forcing the protons sufficiently close together for fusion to take place. (There are other methods for achieving fusion, but they require high-energy particle accelerators.) None of these research programs could be described as simple, everyday, or laboratory-bench scale, as was the method reported by the Utah chemists in March 1989.

Martin Fleischmann and Stanley Pons, with respected careers in electrochemistry, followed a very different line from the megadollar fusion machines. It has long been known that the metal palladium can absorb prodigious quantities of hydrogen. Fleischmann and Pons inserted rods of palladium and platinum into heavy water (water molecules in which deuterium has replaced the regular hydrogen in the conventional H_2O) and switched on an electric current. Fleischmann and Pons thought fusion might occur because of the great concentration of deuterium within the palladium electrodes. They reported observing the release of heat energy at the palladium in quantities far greater than the small amount of electrical energy being supplied. They also saw gamma rays with energy of about 2 million volts as well as neutrons and some of the radioactive hydrogen isotope, tritium (^3H, with a single proton and two neutrons in each nucleus). These are precisely the signals that should accompany fusion. Fleischmann and Pons's experiment not only produced these results but did so with a remarkably modest budget and under normal laboratory conditions—no billion-dollar Tokamaks for magnetic confinement of million-degree plasmas, no giant accelerators. If their claims had been confirmed, they would have discovered a wonderful source of energy. But both their observations and their experimental analysis were rose colored.

Fleischmann and Pons had been working away quietly for several years, using their own funds. Unknown to them, a group of physicists at nearby Brigham Young University (BYU), led by Steven Jones, had also been investigating cold fusion. The BYU group was studying fusion reactions mediated by muons, which occur in nature among the cosmic rays but can be produced under laboratory conditions only in the collisions of very high-energy particles from accelerators. Muon-catalyzed fusion, as the process is known, has been well documented.[10] The Jones group, in collaboration with theoretical physicist Johan Rafelski of the University of Arizona, were also looking into the possibility of cold fusion mediated by an electric current through a liquid. When Jones was asked to review a Utah research proposal sent to the Department of Energy (DOE), he recognized the Utah and BYU teams' common objective and discussed this with DOE. Then Jones contacted the Utah people to suggest the two groups collaborate, as they had expertise in complementary techniques. Full collaboration did not result, but the groups remained in touch and at one time did discuss possible joint or simultaneous publication.

The story then becomes tangled. To some extent, it can be traced through reports in *Nature, Science,* and a host of less scientifically oriented media. It even made the cover of *Time.*[11] The subsequent developments are at least entertaining if not always edifying. Cold fusion itself has followed a Cheshire-cat evolution through press and scientific conferences, congressional hearings, and conflicting results reported from laboratories around the world. Probably the best description of this whole affair has been given by Gary Taubes, who interviewed more than 250 of the people involved.[12]

Fleischmann and Pons's announcement received immediate reaction that ranged from disbelief through skepticism to curiosity. Understandably, because of the well-established nuclear theory, at one extreme there was total incredulity. There was also the response, "Well, perhaps there *is* some totally unsuspected effect that could have great importance." The apparent simplicity of the Utah experiments led to a sort of cold fusion gold rush, with results similar to that of most of the earlier Colorado and California prospectors. Hundreds of scientists dropped whatever they were doing in order to assemble cold fusion kits.

Many scientific discoveries have come after years of designing, building, and laboring with gigantic apparatus or (as with continental drift) from the painstaking accumulation of data from fieldwork before a pattern could be discerned. The published results typically take months to work their way through the editorial process. But with cold fusion the initial results were revealed in a dramatic press conference in Utah on March 23, 1989, followed the next day by a *Wall Street Journal* report. (Palladium stocks went up.) The story had in fact begun to leak earlier, appearing in London's *Financial Times* in the morning of March 23, several hours ahead of the press conference. Thereafter, a technological innovation kept all interested scientists in constant communication. Electronic mail (e-mail) via computer networks carried a scientific samizdat of cold fusion news, gossip, and parody, informing readers of the electronic bulletin boards of every advance (and retreat).

The day after the Utah announcement, the BYU group disclosed their own results, also at a press conference. Jones reported the detection of neutrons—but in numbers a billion times fewer than should have been seen if the Utah energy release were truly from fusion.[13] As the weeks grew into months, the picture became clearer—or hazier, depending on your perspective.

Reports came in from all over. Workers at Harwell, in England, also could not find neutrons in the Utah numbers. A group at the Frascati laboratory near Rome reported positive results, as did people at Georgia Tech, who promptly withdrew their report citing instrumental problems. Scientists at Los Alamos found low-level bursts of neutrons, but none were seen at Yale. Neutron bursts of decreasing size were observed at Nagoya University. Announcements of successful attempts to confirm the tritium detection came from Texas A&M, were withdrawn, and then were re-reported. Among those who claimed success was Boris Deryagin of the Institute of Physical Chemistry in Moscow, who had also been a central figure in the polywater case (Chapter 7).

The mood of the observers can be well gauged from headlines of some of the reports over the following months:

"Cold (Con)fusion"   *Nature*, March 30, 1989
"Fusion Breakthrough?"   *Science*, March 31

"Cold Fusion Causes Frenzy but Lacks Confirmation"	*Nature,* April 6
"Fusion Follow-up: Confusion Abounds"	*Science,* April 7
"Confirmations Heat Up Cold Fusion Prospects"	*Science,* April 14
"Skepticism Grows over Cold Fusion"	*Science,* April 21
"Cold Fusion: What's Going On?"	*Nature,* April 27
"The Utah Fusion Circus"	*New York Times,* April 30
"Science as Spectator Sport"	*Physics World,* May
"Where Fools Rush In"	*New Scientist,* May 13
"Cold Fusion: Bait and Switch"	*Science,* May 19
"End of Cold Fusion in Sight"	*Nature,* July 6
"Cold Fusion: Smoke, Little Light"	*Science,* November 17
"Farewell (not Fondly) to Cold Fusion"	*Nature,* March 29, 1990
"The Embarrassment of Cold Fusion"	*Nature,* March 29
"Gunfight at the Cold Fusion Corral"	*New Scientist,* June 16
"Cold Fusion: Only the Grin Remains"	*Science,* November 9
"Cold Fusion at Texas A&M: Problems, but No Fraud"	*Science,* December 14
"Whatever Happened to Cold Fusion?"	*New Scientist,* January 19, 1991
"High Noon at Utah"	*Science,* January 25

Several workshops or conferences were hurriedly planned to review cold fusion. In May 1989 the DOE appointed a special panel to advise the department on the latest findings and make recommendations regarding research funding. Pons would not provide the panel access to his laboratory without advice from the university's lawyers. By July the panel's draft report noted that "no convincing evidence for cold fusion had been found," and it went on to suggest that no special research funding was justified.

As this drama was being played out in slow motion, there were several side developments. One unfortunate but short-lived portrayal had chemists pitted against physicists in the scientific disagreement, an unfair view of both communities. A small, parallel literature of parody flourished briefly. One tract that appeared on a bulletin board in my department told of an Iowa farmworker who "took a couple of busted cultivator shanks, welded them together and wrapped some electric-fence wire around them and dropped them" into an old bucket filled with antifreeze. His supervisors then forbade him to divulge any further secrets. One of my colleagues summarized a widespread sentiment: "The rumor that the Fleischmann and Pons paper has been rejected by *The Journal of Irreproducible Results* is false." (This journal really does exist.)

Meanwhile, back in the lab, Fleischmann and Pons had submitted a paper to *Nature* then withdrawn it with the comment that they were too busy to attend to the referee's critique. Theorists were not left out of this adventure. Most could not see any reason to expect cold fusion and pointed this out with varying degrees of detail. As might also be expected, a few theorists were able to contrive exotic scenarios to explain the Utah results.

By now, you must surely be wondering, Why was there all of this confusion? Why can't scientists do their experiments, make their calculations, and come to some agreement? What has happened to the vaunted scientific method? The answers to these very reasonable questions can tell us a lot about scientific research.

First, research is *not* always straightforward, despite appearances. Scientists develop laboratory skills that often do not find their way into the published reports. As a result, it is often difficult for newcomers to a field to avoid repeating past errors until they, too, have discovered the tricks. Second, to check on an experiment, one needs a full description of what has been done and seen. Fleischmann and Pons have been extremely secretive. In part, this may be understandable, as they were trying to protect their invention while patent applications were being filed, but this is still far from the scientific norm. Their first paper, in the *Journal of Electroanalytical Chemistry,* was notable for its conversational imprecision.[14] Their continuing lack of openness has been the subject of much criticism.

If the Fleischmann-Pons effect had been genuine and of the magnitude they reported, confirmation *would* have been easy and swift. With almost no exceptions, independent tests very quickly showed that any effects, if they existed at all, were extremely small and close to the limits of detectability, a situation notoriously open to spurious effects and instrumental artifacts. The sensitivity of apparatus to different temperatures, general background radiation, and calibration problems were all the subjects of continuing checks. The statistical analysis of results provided an added area of disagreement. Then there was the question of the accuracy of the gamma-ray energy measurements. An energy of 2.224 million electron volts (MeV) can signal fusion of deuterons, but gamma rays with an energy of 2.204 MeV are emitted by airborne radon, well known as an indoor environmental problem. And so it went. Numerous papers were published describing these experimental problems and carefully setting lower and lower limits to the size of any of the original Utah effects. Fleischmann and Pons retreated, withdrawing much but not all of their initial claims.

Meanwhile (again) there were other developments. The State of Utah very quickly appropriated $5 million to establish an institute for research in cold fusion. Within a month the president of the University of Utah was asking a congressional committee for $25 million. The university announced that an anonymous donor had given $500,000 for the project, but later it was discovered that the "donor" was none other than the university itself. Probably the most surprising development of all took place in May 1990. Faculty in the University of Utah physics department monitored some cold fusion cells in Pons's laboratory and published their results in *Nature.*[15] They could find none of the expected emissions. A few weeks later the physicists were contacted by Pons's lawyer, who demanded retraction of their results to avoid legal action.

In the wake of this extrascientific guerrilla approach, the faculty senate of the University of Utah criticized the university president, who subsequently resigned. The threatened legal action was withdrawn. At around the same time, *Science* reported that there were accusations that the positive results from Texas A&M were not genuine, that the tritium detected came not from fusion but from spiking. This allegation was later found to be without foundation.[16]

Nearly two years after the initial announcement, the *New Scientist* presented two very different summaries of the situation.[17] The first came from Frank Close, an English physicist who divides his time between Oak Ridge National Laboratory and England's Rutherford Laboratory. Close was highly critical of Fleischmann and Pons's efforts and methods. The second, by John Bockris, professor of chemistry at Texas A&M, defended the work of his own and other groups and indulged in some digs at the physicists. (Some years later, Bockris became embroiled in another controversy when he undertook experiments to transmute nuclei of base metals into gold. A number of his departmental colleagues did not appreciate this latter-day alchemy and signed a petition urging him to resign.)

To some staunch defenders, the issue, as they say, is unresolved, but most other scientists are unpersuaded of any real fusion. By June 1991 there was the announcement of the closing of the Cold Fusion Research Institute that the State of Utah had set up. Over the next few years, there were sporadic reports of further research and additional claims for signs of fusion, but nothing has yet been able to restore major credibility to the early claims. In January 1992 a cold fusion researcher at SRI International was killed in a laboratory explosion. In December 1993 the University of Utah sold its commercial rights to cold fusion for $600,000 to Eneco and is to receive royalties should any commercial success occur. Fleischmann and Pons continued to pursue cold fusion and authored a lengthy paper, "Possible Theories of Cold Fusion," in 1994.[18]

A chronicle of the cold fusion fiasco has been the subject of several books; the first, by Frank Close, contains an especially devastating description of how Fleischmann and Pons first reported a gamma-ray signal at an energy of 2.5 MeV.[19] After it was pointed out that 2.2 MeV was the value expected from fusion, the same data reappeared with a shift of scale so that the peak was now at the correct place. The literature on cold fusion continues to grow. Taubes's book has already been mentioned. John Huizenga, chairman of the National Academy of Sciences committee that reviewed the whole case, has written a highly critical account. In contrast, Eugene Mallove, formerly a publicist at MIT, has written in support of the reality of cold fusion. Finally, at least for the moment, there is an extensive review by Bruce Lewenstein in *Osiris*, a journal of the History of Science Society.[20]

To summarize briefly, the cold fusion "discovery" will surely be remembered as a striking example of how science should *not* be done. Taubes has compared "many of the proponents of cold fusion" to Blaise Pascal, the seventeenth-century scientist who "renounced a life of science for one of faith."[21] The whole episode certainly illustrates the practical difficulty in implementing an innocuous-sounding "replication" and points to the need for full and open disclosure if there are to be meaningful tests and checks. It has also exposed some unfortunate professional sensitivities, jealousies, and resentments. At least to date, the exercise appears to be devoid of redeeming scientific value—but perhaps something may yet turn up as the few holdouts tenaciously pursue a theory as evasive as the Cheshire cat.

Notes

1. H. E. Le Grand has provided an excellent description with insightful analysis. See his *Drifting Continents and Shifting Theories* (Cambridge: Cambridge University Press, 1988).

2. *Principia Mathematica,* ed. Florian Cajori (Berkeley: University of California Press, 1962), 398.

3. *Nature* 111 (1923): 226.

4. *The Earth* (Cambridge: Cambridge University Press, 1924), 261.

5. Quoted in W. van der Gracht, *Theory of Continental Drift: A Symposium* (Tulsa, Okla.: American Association of Petroleum, 1928), 145.

6. *American Journal of Science* 242 (1944): 218, 514; 248 (1950): 584.

7. *Drifting Continents,* 229.

8. *New Scientist* 86 (1980): 234.

9. *Science* 246 (1989): 575.

10. *Scientific American* 257 (July 1987): 84.

11. May 8, 1989.

12. *Bad Science: The Short Life and Weird Times of Cold Fusion* (New York: Random House, 1993).

13. *Nature* 338 (1989): 737; see also 711.

14. M. Fleischmann, S. Pons, and M. Hawkins, *Journal of Electroanalytical Chemistry* 261 (1989): 301.

15. *Nature* 344 (1990): 401.

16. *Science* 248 (1990): 1299, and 250 (1990): 1507.

17. *New Scientist* 129 (January 19, 1991): 46.

18. *Il Nuovo Cimento* 107 (1994): 143.

19. *Too Hot to Handle* (Princeton: Princeton University Press, 1991).

20. John Huizenga, *Cold Fusion: The Scientific Fiasco of the Century* (Rochester, N.Y.: University of Rochester Press, 1992); Eugene Mallove, *Fire from Ice* (New York: John Wiley and Sons, 1991); *Osiris* 7 (1992): 135.

21. *Bad Science,* 92.

4

Science and Its Practice

While I have been describing some of the bizarre penumbral science that has often attracted a great deal of attention, I have avoided any detailed examination of the question of how the scientific contents of those episodes are to be judged. What are the standards against which they are measured? Who is qualified to pass judgment on the validity of the claims? Indeed, and most fundamentally, what *is* science? What is the *scientific method* so often invoked as though it were some magical procedure?

In the generally held picture of the scientific method, there is an endless cycle of experiment and theory. "Experiments" actually consist of carefully designed and controlled experiments and observations of naturally occurring phenomena. From these experiments and observations, broad generalizations are extracted (by induction) and synthesized into theories to "explain" what has been seen. We can put the theory to the test by making more measurements. To the extent that we get agreement with the formula's predictions, we will continue to use it with confidence. But if we find disagreement between prediction and experiment, we may have to modify the theory or even seek something totally new that rests on different assumptions.

This brief sketch of the scientific method is correct as far as it goes—but it omits any number of important factors. What sort of assumptions are acceptable in constructing a theory? What constitutes a satisfactory scientific theory? How do scientists come to accept some theories but reject others? Who makes these decisions?

Over the years these questions have been answered many times, often at great length but without finality. Commentators (like blind people feeling an elephant) tend to describe different aspects; the whole is far more than the sum of the parts. Most scientists have probably not given much thought to these questions. Ask a scientist to define science and the scientific method, and you will probably get a description that is correct as far as it goes but most likely very incomplete. Ask a philosopher of science, and you will probably receive a much broader answer that will often seem (to many of us scientists) to make only occasional contact with what we recognize as our activities. In addition to these familiar views of the scientific enterprise, another perspective has been emerging from the the the sociologists of science. The social dimension covers the interactions within the scientific community as well as the complex relations connecting science to the wider society.

What are these different views of science? Most working scientists, if asked, would probably give a "realist" opinion: that there exist in nature a number of regularities that relate various entities and processes. The task of the scientist is to discover these regularities and express them as "laws," often using the language of mathematics. Later scientific research may point to the errors or incompleteness of an existing theory and lead to its modification or even its rejection in favor of a more comprehensive theory. The test of a theory lies both in its ability to draw together a broader range of observations than did its predecessor as well as its ability to predict the outcome of future experiments and observations. In the realist view the content of science is cumulative, apart from obvious error. The basic experiments and observations are "facts" that remain true, though their incorporation into theories may change with time, generally gradually but sometimes radically. To that extent the core knowledge of science is reliable.

In contrast, there are strenuous debates among philosophers and sociologists of science regarding the structure of science. One school of thought plays down the role of experiment in the construction and validation of theories to such an extent that, in rebuttal, Allan Franklin felt it necessary to argue that "it is reasonable for scientists to gather data."[1] Even the reality of the success of science has come under scrutiny. A view of science currently held by some sociologists and philosophers is termed "relativist"; it is based on the assumption that scientific knowledge is relative, dependent on the views and social ambience of each commentator and without any underlying physical reality. Consequently, much as social and political views change with time and vary between different groups, so will "scientific" interpretations change. Further, in the words of Stephen Cole, "It is also clear that much of what was commonly accepted by the scientific community as true in the past is currently believed to be wrong. What we currently believe to be true *will* in the future most likely be thought of as wrong" (emphasis added).[2] The implication of this view is that no scientific knowledge can be trusted, or, as John Ziman has described it, "doctrinaire sociologist relativists seem to suggest that one bit of claimed knowledge is as good as another. ... Scientists have every right to express their opinion that some knowledge claims, such as those made for extrasensory perception, are so contrary to established understanding, and are supported by so little evidence, that they should be dismissed as parascience."[3]

There is no way in which I can synthesize all of these ways of viewing and describing science into a single comprehensive picture that will satisfy everyone. Nor can I completely evade a discussion of these topics. I therefore give emphasis to the image of science that most scientists see and set out my own descriptions, leaning heavily on the thoughtful writings of Ziman and Thomas Kuhn, which I have found most congenial to my own perceptions. In doing this, I am well aware that I will attract, at best, the scorn of the relativists, but I am not alone in my views. Larry Laudan, a philosopher of science, has attacked "the displacement of the idea that facts and evidence matter by the idea that everything boils down to subjective interests and perspectives," which he calls "the most prominent and pernicious manifestation of anti-intellectualism in our time"; he believes "the relativist position to be profoundly wrong-headed."[4] Another trenchant criticism of the "new sociology of science" has come in two lengthy papers from the scientist-philosopher Mario Bunge, who writes that the result is an "utterly grotesque picture of science."[5]

In the main, scientists are likely to agree with my descriptions, and it is scientists whose opinions are going to determine what is welcomed and then incorporated as new components of science or be rejected as erroneous or pseudoscientific. In order to understand scientists' reception of unorthodox ideas, we need to understand science as scientists know and use it, not as others might see it or prescribe procedures for it.

In defense of the realist position, let me cite just two of the clear successes of scientific prediction. After the accidental discovery of the planet Uranus in 1781, careful observations showed that its orbit deviated slightly from what was predicted. Those calculations took into account the gravitational attraction of the sun and other planets. Two mathematicians, John Couch Adams in England and Urbain Leverrier in France, took those orbital deviations to indicate the presence of a hitherto unsuspected additional planet. In 1846, astronomers at the Berlin Observatory found this planet very close to its predicted position, and we now know this body as Neptune. More recently, in the 1980s, a complex theory of what are termed "elementary particles" required the existence of a particle that had never been observed. Two groups of experimenters, using totally different methods, found this particle (now labeled the J/Ψ particle).

I find it difficult to accept the idea that Neptune and the J/Ψ particle are simply "social constructs." Against this sort of success, we have, as far as I know, no comparable and testable prediction from the relativists. Certainly nothing has emerged that is of any help to us in determining the scientific merits of claims for new discoveries, nothing to evaluate the merits of cold fusion or continental drift or the role of DNA in biological structures and processes.

Kuhn's book *The Structure of Scientific Revolutions* appeared in 1962 and has become the most widely read and influential book of its type over the past fifty years.[6] Although Kuhn's views have since changed somewhat, his book has in many ways shaped the debates over the descriptions of the methods of science and, more than any other work, has caught the imagination of scientists who have generally been at best skeptical of the philosophy of science and even less complimentary about the sociology of science. (But in honesty I would guess that most scientists have probably not read Kuhn.) I have also found Ziman's *Public Knowl-*

edge[7] of great value in setting out a description of the scientific enterprise, covering many aspects that Kuhn did not touch upon but that are equally important and extend into the sociology of science—or what is now becoming known as "science studies," the social structure and dimensions of science. I have no doubt that the reason I find their writings so accurate as descriptions of science is that both authors were trained as scientists who tend to approach this subject as I would. Kuhn was involved in physics research before switching to the history and philosophy of science, and Ziman had a distinguished career as a theoretical physicist whose writings on the structure of science led him into science studies. Another scientist whose writings I have found perceptive and useful is Peter Medawar, 1960 Nobel Prize winner in physiology.[8]

Science represents the human effort to describe and understand the natural world through passive observations, active experiments, and theoretical analysis and synthesis. If we judge a science and its theories by the simple test of their usefulness for making accurate and generally quantitative predictions, we find, in the natural sciences and especially in the physical sciences, that mathematics has become an indispensable language for describing the relationships between various quantities and for permitting the manipulation of those relationships to yield insights and predictions that would be totally impossible if words alone were used. When evaluated in this way, the physical sciences have been the most successful sciences largely because it has been possible to reduce their complexity to the scale of controllable and inanimate systems, with the focus of attention on a very few quantities to the exclusion of all others. In the biological and medical sciences, it is not as easy to identify much less exclude unwanted variables in the study of living organisms, but the application of molecular and cell methods has produced a revolution in some areas. Moving along the spectrum of success, we find some problems in the social and behavioral sciences that have yielded to the application of quantitative methods, but many more that elude even agreement on methods, let alone the meaning of their results. Perhaps very different scientific methods will have to be devised and our definitions of science thus expanded.

A foundation for success in a mature science is the widespread acceptance of a number of basic assumptions and definitions, as well as the methods by which activities and their results will be judged. In the absence of this professional consensus, opposing camps find it difficult to conduct fruitful debates and therefore to accept each other's findings. This doctrinal division can have practical consequences. For example, in psychiatry, one school of thought relies heavily on the physics and chemistry of the body for its diagnoses and treatment, while another school seeks subconscious causes for behavioral dysfunction. In contrast, consensus is overwhelming in the physical sciences: There is virtual unanimity on the foundations, still leaving plenty of room for disagreements along the frontiers. Physicists *do* explore possible small modifications to Newton's laws, and evolutionary biologists *are* involved in a vigorous debate on the speed and pathways of evolution. At the same time, we do not find physicists haggling over the major framework that includes such sweeping principles as the conservation of energy or biologists questioning the central idea of evolution. Both of these have long since passed into the well-established body of accepted science, and it is their fine-tuning and expansions that attract attention.

Accordingly, as scientists, most of us go about our daily work comfortably within the accepted norms of our particular speciality, generally untroubled by skirmishes being fought out over confessional differences. What, then, guides us in our day-to-day science? We do not start each day by using the flip of a coin to decide what to do, which experiments or calculations to pursue. In our "normal science" (as Kuhn has termed it) we work on *problems* whose articulation is drawn directly from the prevailing theory. To this structure Kuhn has applied the label *paradigm*, intended to encompass both the experimental methods and theoretical framework, a consolidation of past success that defines a program for further research. A paradigm can be as broad as the theories of evolution and plate tectonics or as localized as the theory of radioactivity. A *mature* science (to use another of Kuhn's terms) has progressed far beyond the early exploratory stages and in its self-confident stability has paradigms that are widely shared and guide the choice of further research projects. We choose our next problem for attack because it makes sense under the umbrella of the prevailing paradigm. We promptly reject ideas that fail to fit in unless there is a good a priori case for their plausibility. Thus, accepting Velikovsky's theory would have required us to believe in the occurrence of events that ran counter to so many successful uses of the mechanical quantities of force, momentum, and energy that this did not seem (to most scientists) to be worth pursuing. That negative view could, in principle, have been changed if Velikovsky had produced plausible evidence that our mechanical paradigms were at fault or incomplete, but this he failed to do. Velikovsky's supporters, almost exclusively untrained and certainly ill informed in physics, did not accept our paradigms or seemed to be willing to accept capricious relaxations in our laws of motion and provident pathologies in the behavior of well-known forces. There is a difference between having a mind that is open to new ideas and one that is simply vacant. In contrast, when Einstein challenged those same mechanical ideas, he *did* force a change in the ways in which force, energy, and momentum are defined and used. He did this by showing how James Clark Maxwell's equations of electromagnetic theory (a cornerstone of one of the nineteenth century's most successful theories) had certain shortcomings that could be remedied only through some radical assumptions that then had testable implications. A few years later he did the same with Newton's law of gravity. Einstein's theories have subsequently been confirmed repeatedly and with great accuracy.

Most scientists do not think that Newton was wrong but rather that his laws relate to a narrower range of situations than do Einstein's. Although Einstein's formulation rests on a view of time and space that is radically different from Newton's, mathematically it is easily shown that Newton's formulations are contained within Einstein's. For practical purposes, it is often much easier to use Newton's version, and this is routinely done for most space probes and satellites.

Within a prevailing paradigm, there can be many different reasons to choose a research problem. We might just be curious—what will be the outcome of this calculation or that experiment? This curiosity is generally not entirely innocent. We might wish to measure a quantity because of its potential usefulness in other calculations or experiments. For example, we might want to measure the behavior of some plastic under the influence of solar ultraviolet rays because we are trying to make biodegradable materials. We don't plan a fundamental challenge to a paradigm—applied research is the stuff of industrial success. Or we might undertake a calculation of some quantity that will determine whether an experiment needs to be redesigned to increase its sensitivity. Or we might be planning to measure the region of validity of a theory—does it work at those speeds or at this temperature or under zero gravity or whatever?

In selecting research problems, we are strongly influenced by our impression of the solvability of the problem. As Medawar has pointed out, we are very unlikely to embark on a research project if we do not even know whether an answer, any answer, exists. This may require us to make quick preliminary measurements or calculations to give enough hope that we will learn something from a larger investment of time and resources. Modern science, in Medawar's opinion, has made its great progress by selecting just those problems that *can* be solved and ignoring problems that seem interesting but for which there seems to be no solution with current methods, if at all. In contrast, in the social sciences many problems are obvious because of their great social importance, but they do not necessarily have unambiguous solutions, solutions that do not depend on subjective and unvalidated assumptions. A social problem may still need a solution, but this will not be in the sense of solving the problem but rather of dealing with it by finding a political or social compromise (at least temporarily).

We work within the framework of a prevailing paradigm. How did it emerge? It was assembled from the accumulation of earlier generations of experiment and theory. In some cases *induction* was used: From a number of specific examples a broad generalization was drawn that describes not only those measurements that have been made but also those that have not yet been made. For example, Galileo determined that the distance an object falls increases with the square of the time taken. He did not measure every possible distance and the time for each corresponding fall; his results, though, have been generalized into the formula

$$(distance) \quad is\ proportional\ to \quad (time)^2.$$

This result can be used to calculate the time taken for *any* specified distance or, in the inverse calculation, how far an object will fall in a given time. This formula works even for distances not originally measured.

We must refrain, though, from thinking that induction is the only way to scientific progress. Generalizations do not always follow from a hunting-and-gathering phase. One does not assemble a winning baseball team by scanning a selection of baseball cards. Many of the most revolutionary advances have come from flashes of inspiration and intuition, by processes not at all well understood. The same can be said of the assembly of a pennant-winning baseball team.

Why doesn't the formulation of a paradigm close out research in that subject? What need is there for further research? Why test a paradigm? There are several reasons. Take again the example of Galileo and the falling object. We might be interested in testing how far this formula can be extended. We have no conclusive evidence that Galileo actually dropped anything from the tower in Pisa; we do know that he devised some ingenious ways of converting the problem of the falling object into one that allowed him to make useful measurements with the relatively crude timing means at his disposal. But even if he had gone to the top of the tower, he could have conducted his tests on objects falling over distances of no more than the height of the tower, about 200 ft. Does the same formula apply to objects falling 500 ft, 1,000 ft, or more? Here we are trying to *extrapolate,* to go beyond the known region of the paradigm's validity. Experiments show that our simple formula indeed does *not* hold indefinitely, and we have come to understand this in terms of air resistance that builds up as the speed of an object increases. At low speeds this effect is too small to be noticed, but if we plan to make calculations for very large distances of fall, then the theory and resulting formula need to be altered. Even without air resistance, the simple laws do not hold indefinitely. At very high speeds it is necessary to use Einstein's relativity theory. The familiar low-speed laws are then seen as contained within the formulas of relativity. For mathematical convenience, though, we continue to use the old familiar formulas for many everyday computations—even in such cases as the trajectories of space probes—because the accuracy is quite sufficient.

There are additional reasons a paradigm does not represent the last word. In the course of normal science, we might find results that are unexpected, that do not seem to find an explanation in terms of the paradigm. Kuhn terms these findings *anomalies.* When we encounter an anomaly, our immediate response should be to check measurements and calculations. It is very easy to misread a dial or make an error in a calculation, and we have all done so. Sometimes our apparatus produces an artifact—something that we never satisfactorily explain but that never recurs. Reports of some of these anomalies do find their way into the scientific literature, where they may play a useful role, stimulating other scientists to be alert to possible new phenomena and so trying to confirm their existence. But many of these anomalies remain unconfirmed and unexplained; this aspect of the published literature is known to the "in" groups but stands as a potential trap for the unwary. Not everything in print is reliable, as we should know so well from newspapers. Science is not immune to this problem.

Alternatively, we might accidentally stumble across something qualitatively new, some phenomenon never before observed. Again, we check our measurements and calculations. In some circumstances it is not necessary to have more than a single example to establish the reality of the interpretation. In other cases new effects are discovered only after the accumulation of enormous collections of data and an analysis that shows the results to be statistically different from what was expected. When I was a graduate student, our laboratory found several new types of subnuclear particles. There were instances where straightforward explanations of single instances were so clear-cut that there could be no doubt that we had found new types of particles; there was no plausible alternative explanation, no need to wait to accumulate hundreds or thousands of confirming examples. Colleagues found other particles, equally unexpected.

Pseudoscientific claims are often rejected because they have no theoretical foundation. This objection is not always valid. Our particle discoveries were totally independent of any theory. There was no theory, no paradigm to guide us. As new fields open up through accidental discoveries, there may be no theory to support them. It is the reality and correctness of the observations that must be examined, and the theory will follow in due course if the observations are correct. The reality of continental drift does not depend on the existence of a theory to explain the movement of the continents, though the acceptance of the idea was certainly slowed by the apparent power of Jeffreys's objections, based on his calculations and reputation. That these objections were later seen to be misleading did not detract from their influence at the time. Well-founded theory contradicting an idea or even a lack of theory to support it may make us skeptical, with good reason, but may not provide a sufficient basis for rejection of the idea. The issue is the relevance of whatever theory there is, as with Jeffreys, although its limitations might not be seen at the time. No theory means simply that: no theory. This is very different from a situation in which the existing theory suggests very strongly that the new claim (such as for cold fusion) is most probably in error.

When anomalies or accidental discoveries turn up, we try to obtain confirmation. We therefore expect the published reports to have sufficient detail to tell us just what has been done. In principle this will allow us to try to repeat the experiment or calculation. The simple version of reproducibility is the ability in principle for someone else to build the needed apparatus, repeat the experiment, and obtain the same results or to undertake the same calculations. This is not quite as obvious as it looks. A calculation can involve mathematical or computational shortcuts whose choice can influence the results in subtle ways. Reproducibility has been considered a critical test for scientific discoveries, but I have added the qualification "in principle," for reproducibility may no longer be possible in many areas of modern science. When a result comes from a collaborative effort of more than 100 scientists, using, for example, a gigantic machine or space probe, there may be no way of attempting to reproduce the experiment. The human genome project is so large that independent confirmation of its findings is probably not practical. Some computer calculations are so complex that it is an academic question to expect an exact repetition as a check. The best we can do in many circumstances may be to undertake a complementary experiment or calculation, using a different machine or program, and look for consistency between the results. (But this assumes that we know enough to make such a comparison, and that in turn implies the existence of a satisfactory theory.) Strict reproducibility can no longer be demanded in all cases.

What sorts of tests can we apply to decide whether any particular result is close to what our paradigm predicts or lies well outside an acceptable range and so qualifies as an anomaly? In some cases there may be no paradigm, but the quality of the result is unambiguous. Thus, when we found a subnuclear particle that was radioactive and produced an electron, and nobody had ever before seen an electron under these circumstances, the conclusion was clear: We had discovered a new kind of particle. When a coelacanth was caught off the southeast coast of South Africa in 1938, there was no question as to its unique nature, an example of a fish that had long been thought to be extinct. There was no need to await the discovery of a second example of this remarkable fish.[9] But the identification of many anomalies rests on statistical evidence to answer the question, Is it clear that this result is higher or lower, bigger or smaller than we had expected? In such cases we are making a comparison between measured and expected values, each known only to within some specified accuracy, and we ask whether the two values (observed and expected) differ *significantly*.

All experiments are open to error, and we identify two types. *Systematic* errors are those which shift a measured value in one direction, higher or lower depending on the circumstances. For example, our measuring scale might have been wrongly calibrated, so that every value is systematically 1 percent too high or perhaps 2 percent too low. Once we identify and measure this systematic effect, correction is easy, but the truly difficult part is even knowing that we have a systematic effect present. We try to guard against this type of error by careful calibration, by making measurements under varying conditions, and, where necessary, even by using different people to guard against observer error, well known where visual observations are still made, as in astronomy or in microscope work in the laboratory.

There can also be statistical errors leading to measurements that are more or less equally spread above and below the true value. If we make a graph to plot all the measurements, we often get the familiar bell-shaped curve that peaks around the true value and shows decreasing numbers of measurements as we move further away from the peak. The spread in the measured values is a measure of the reliability that we can attach to the calculated average. We can then quote our answer as $x \pm y$, meaning that x is our best estimate of the true value and y, the measure of uncertainty, is a *standard deviation*. From the well-explored theory of experimental errors, we expect that there is about a 68 percent probability that the true value lies within the range from $x - y$ to $x + y$; a 95 percent probability that it lies within twice that range, from $x - 2y$ to $x + 2y$; and a probability of only 0.3 percent that the true value lies outside the three standard deviation range, $x - 3y$ to $x + 3y$. We talk of levels of significance and take very seriously disagreements that rest on differences of more than three standard deviations. To reduce the standard deviation, we can increase the number of measurements or try to refine the experimental technique either through improved procedures or even by designing new apparatus. The improvements we can make in our hardware, though, are limited by the basic properties of the materials and the inexorable laws of optics, electricity, and so on.

In practice, statistical analysis is more complicated than my simple introduction has described. There are experimental situations where the distribution of measurements does not follow the bell-shaped curve, but many variants and their mathematical formulations are well established. There are also practical limits to the improved accuracy we can attain by simply making more measurements, for the improvements increase more slowly than the number of measurements. How many lab mice can we use for a drug test for some specified accuracy? Perhaps 1,000. To improve the accuracy by a further factor of ten, we would need not 10,000 but 100,000 mice. Is this likely? Cost is usually proportional to the scale of the experiment. At some stage we reach a limit set by cost and practicability. We may thus *never* know the values of some quantities to within the accuracy that we would like to have.

Alvin Weinberg invented the term *trans-scientific* to describe this sort of situation, and it is of more than academic interest.[10] We find it central to many environmental problems where we wish to know the effects of some very low-level insult. There can be great political pressure for legislative or administrative action. What *is* the safe level for radiation or pesticide residue or whatever? Scientists may not be able to solve such problems, which require instead a balance between environmental hazards and political and commercial costs. At best, in such cases, science can only identify the scale of our ignorance, but in the end regulatory agencies will have to prescribe exposure guidelines even in the absence of a clear scientific answer.

Experimental scientists often do "feel" when an experiment is behaving and when it is not, but there is a great risk of bias when this "feeling" is used to justify throwing out some of the data: Should this measurement be thrown out because you forgot to make a calibration check (even though the value itself looks good), or should that measurement be rejected because it is so far from the average and there are indications that the temperature went too high during that run? As discussed in Chapter 10, in recent years it has been found that in his precision experiment to measure the electric charge on the electron, Robert Millikan did not report all of his data; his notebooks include much material that seems in disagreement with his main results and that he accordingly omitted from his published manuscripts. His notebooks are peppered with comments such as "Error high, will not use" and "Might omit because discrepancy."[11] Rigorous experimental planning can obviate such problems by setting out a protocol *before* any measurements are made and deciding just what the acceptance criteria should be. In the end we should take the statistical uncertainty as a guide and not as an enforceable warranty to the reliability of our answer. We have, in summary, a reasonable but not perfect procedure for identifying anomalies.

During the course of normal science, the anomalies will accumulate, some to be ignored (rightly or wrongly), others to become the focus of further attention. It is a matter for fine judgment, a matter of experience and intuition, which anomalies to ignore and which to pursue. When sufficient anomalies have been assembled, a pattern might emerge, and the scientists in that field will have to begin the reexamination of the paradigm. Among the popular misconceptions of science is the idea that a theory collapses when confronted by a single "fact" that it cannot explain. Things are not quite so simple. Perhaps the theory can be refined in some minor way to accommodate the anomalies. This is often a satisfactory outcome, because the familiarity of the paradigm and the range of its previous successes are valid reasons for not dumping it without very good cause. If there is a competing theory, we may be more willing to switch allegiances, but often there is no ready alternative. We may recognize the inadequacy of a theory long before any substitute has appeared.

Challenges to a theory come as anomalies increase in number or significance. The scientific "facts" that constitute the anomalies and the theories that are current do not gain acceptance through any formal mechanism. There is no forum where the scientific majority votes its approval; no single scientist acts as a decisive authority; no bench of scientific judges pronounces that the facts have been established beyond a reasonable doubt and then passes sentence on the theory. The verdict emerges in Quaker fashion, by consensus among the knowledgeable professionals.

Gradually or suddenly, the prevailing opinion accepts the reality of the anomalies, and the theory is in trouble. This procedure is the inverse of the popular view that theories are proved or confirmed by successful tests. Such tests do give us added confidence in the theory but never with finality. It is the anomalies that are potentially more powerful, for they can *falsify* a theory. The idea of falsifiability was introduced by Karl Popper, a philosopher of science. Among the points he made in setting out this idea were:

- Every good scientific theory is a prohibition: it forbids certain things to happen. The more a theory forbids, the better it is.
- A theory which is not refutable by any conceivable event is non-scientific. Irrefutability is not a virtue of a theory (as people often think) but a vice.
- Every genuine test of a theory is an attempt to falsify it, or to refute it. Testability is falsifiability. ...
- One can sum up all of this by saying that the criterion of the scientific status of a theory is its falsifiability, or refutability, or testability.[12]

The criterion of falsifiability was applied with great force in the legal dispute over creation science, described in Chapter 11.

Sometimes, despite all the salvage efforts, a paradigm is leaking badly and cannot be patched up. This "crisis" phase (Kuhn's term) may not be quickly resolved. The fifty years of debate on continental drift showed how slowly the anomalies can come to be accepted. In comparison, there was a much shorter crisis stage in the exploration of the properties of certain mesons, subnuclear particles that were discovered among the cosmic rays before they could be manufactured under laboratory conditions in the collisions of very energetic nuclei. In certain respects the behavior of the mesons was not in accordance with some cherished theoretical principles. We checked our measurements, the theorists wriggled in their calculations, but the anomalies persisted. Finally, after about five years, T. D. Lee and C. N. Yang in 1956 put forward what was considered a revolutionary idea. Within a year several experiments had shown them to be correct, and they received the 1957 Nobel Prize.

Similar drastic changes in the scientific ways we view the world accompanied the introduction of the theories of relativity, quantum mechanics, and evolution, to name just a few. Many revolutionary changes will be perceived only by specialists, but evolution, relativity, and quantum theory have implications that are so widely known (or misunderstood) that they influence our thinking, our culture, and even (as with evolution) some legislation.

What is worth noting is that the progress of each revolution generates a momentum of its own. The pace at which evidence accumulates (in the form of anomalies) varies widely. There comes a stage where adoption of the new paradigm becomes inevitable. The professional consensus changes from skepticism to overwhelming acceptance, and we enter another period of normal science, guided now by the new paradigm.

One final point remains to be made. As new "facts" are reported and claims are made in support of or opposition to an existing theory, the initial burden of proof rests with the scientist making the claims. A typical pseudoscientific tactic is to attempt to reverse this, to present some revolutionary claim with the implicit challenge, "Prove me wrong." Science does not work this way. It is up to the challenger to make a plausible if not yet compelling case. Only then will the scientific community dignify the new ideas with a reasoned response. Even a quick check will show, to an experienced scientist, the weakness of some new "idea." Most of us have too many other tasks to be willing to divert our efforts on some clearly wrong proposal, even if sincerely presented. Scientific journals are not going to dilute their contents with the wild claims and speculations that arrive with depressing regularity. Some are modestly presented, some arrogantly, and some even accompanied with threats. None of this will affect our scientific judgments.

The scientific method, as it has evolved, has been remarkably successful. In its conservatism it has constructed filters to keep out the cranks and probably made it somewhat more difficult for genuinely scientific but radical ideas to be aired.

But as I show in Chapter 6, the nature of many ideas that science *has* entertained is often far more radical than those it peremptorily rejects.

The structure that I have outlined in this chapter is the theoretical framework of the scientific method; the apparatus that makes it work is the subject of the coming chapter.

Notes

1. *Experiment, Right or Wrong* (Cambridge: Cambridge University Press, 1990), 102. Franklin is a physicist who has applied his experience in his perceptive writings on scientific method.

2. Stephen Cole, in *Making Science* (Cambridge: Harvard University Press, 1992), 11.

3. *An Introduction to Science Studies* (Cambridge: Cambridge University Press, 1984), 187.

4. *Science and Relativism* (Chicago: University of Chicago Press, 1990), 10, 11.

5. *Philosophy of the Social Sciences* 21 (1991): 524, and 22 (1992): 46.

6. *The Structure of Scientific Revolutions* (Chicago: University of Chicago Press, 1962, 1970).

7. *Public Knowledge* (Cambridge: Cambridge University Press, 1968).

8. *The Art of the Soluble* (Middlesex, England: Penguin, 1969), and *Induction and Intuition in Scientific Thought* (London: Methuen, 1969).

9. For more on the coelacanth, see K. S. Thomson, *Living Fossil: The Story of the Coelacanth* (New York: Norton, 1991).

10. *Minerva* (April 1972): 209.

11. Cited in Gerald Holton, *Historical Studies in the Physical Science* 9 (1978): 161.

12. *Conjectures and Refutations* (New York: Harper and Row, 1963), 36.

5

The Machinery of Science

In the preceding chapter I gave an outline of a conceptual framework for science—how findings are assembled into an overall picture from which general formulas or theories or paradigms can be distilled and tested. These paradigms then serve as a framework to guide subsequent research. Scientists will probably recognize and agree with most of my description, but there will be dissenting opinions among the philosophers and sociologists of science. These differences in perception will not be resolved easily if at all, though there will probably be a greater measure of agreement on the topics that I discuss in this chapter: the machinery of science and the ways in which the scientific community shares its findings and speculations toward its goal of new and expanded understanding through consensus.

What are the components that make up the structure of science? In what follows I describe the methods of communication, formal and informal, through conferences, journals, and personal contacts. I also discuss the roles professional societies play, and I include in my survey the education of scientists because of the influence it has in shaping our conception of our trade. Although this coverage might appear to range far wider than the scientific fringes, it is needed. For this scientific machinery is responsible for much of the progress that science is now making, but the same machinery can serve to filter out unconventional ideas. Indeed, it is just the conventional operation of the machinery that has been accused of acting as a censor, protecting the scientific establishment from the challenges of those who are not "in the club" but to whom (they claim) new and often revolutionary truths have been revealed.

Professional journals constitute our most important avenue of communication. Through our journals we report in detail our results and ideas, and we learn of the work of others. We may also share our information with some colleagues through informal networks. In the early days of the Royal Society, Henry Oldenburg exchanged letters with the leading scientists of Europe, sending his correspondence via friends. Then came the regular postal services, and we now employ the much faster but less glamorous e-mail and fax. Journals, though, are irreplaceable. We continue to rely on the printed page (and its modern microalternatives) for an archival role and the screening of their contents by editors and referees.

The frequency with which journals appear covers a wide range. *Nature, Science,* and the *Physical Review Letters* appear weekly; the *Journal of Biological Chemistry* brings us 600 pages every ten days; and the *Journal of Physiology* adds 700 pages to our shelves each month. There are also yearly publications, such as the *Annual Reviews of Earth and Planetary Science,* that contain survey articles of selected topics. All of these publications, regardless of their frequency of appearance, are aimed at fairly narrow professional groups. In them we find the original contributions at the highest professional levels. Their style is jargon laden, and they presume an extensive background on the part of their readers. Other publications (described later) cater to more general readership. *Nature* and *Science* are unusual in having general news and comment sections in addition to those that carry reports of original science.

Editors solicit review or survey articles, guided by the state of each field: Is this a good time for a comprehensive article on subject A, or is it many years since we covered subject B? These reviews are extremely useful; they constitute the major secondary literature, entry points into subjects with which we are not familiar, without the need to have to search out the primary literature and decide which of those papers are important or have been superseded or can safely be ignored. A good survey will, by its selections, omissions, and critical comments, come closest (in some ways) to reviews in the arts and literature, in expressing evaluations of the merits of the published materials. As we will see, an important feature of such journals of current science is that the critical review takes place before publication. In contrast, in the arts and literature, postperformance reviews are the norm. A few journals, such as *Statistical Science,* carry extensive responses and a rebuttal directly following an original contribution. Of course publishers of nonscientific books do solicit manuscript reviews, but the postpublication reviews can often be critical to commercial success. In the sciences book reviews do not assume anywhere near the importance they have in literature; they do not constitute a well-defined art form nor do they usually figure in one's list of publications. These reviews do, though, serve a very useful publicity function, considering the paucity of displays of books on science in many bookstores.

Valuable though the review articles are, it is the primary literature that is the most important. Here we find experiments described, their results analyzed, theories proposed and taken apart, predictions made and tested. Depending on the style of the journal, more or less technical detail is allowed—the optical properties of some material, the chemical preparation of some new compound, the tricks and safeguards that were necessary in some complex computation. The literary style of these technical papers (if that is not an oxymoron) is usually clear (to the experts) but scarcely evocative; written in the third person, passive voice, they tend to be dull. (It is generally considered bad form to use the first person in scientific treatises, though some scientists become known for their confident, first-person descriptions.)

The sociologists Trevor Pinch and Harry Collins neatly described the style of the scientific paper:

> The language is self-effacing, suggesting that the experimenter played the role of facilitator, or amplifier of Nature's voice. ... First results for some new effect are presented modestly, with a provisional air. ... In times of controversy the formal literature becomes conspicuous for what it does not include. Biographical details of the authors are not to be found [nor are] details of the experimenter's health, ... the motives, interests and emotions of the experimenter, ... family and time pressures.[1]

But no matter what the style, the level of detail, there is one underlying component that all of these publications use, and that is peer review. Peer review is the editorial scrutiny that all scientific manuscripts must pass before they appear in print. Each manuscript that is submitted is sent by the editor to two or more referees for their comments. Only in this way can a journal editor be assured that a manuscript is reviewed by fully knowledgeable scientists, for no editor can be an expert in all topics. For most journals, the identification of these referees is not revealed to the authors. Referees are not paid for their services but undertake this often very burdensome task as a professional obligation. Behind their screen of anonymity, the referees are free to express with candor their evaluations of the manuscripts, uninhibited by professional friendships. Good referees can improve a manuscript, suggesting additional measurements or calculations, spotting weak points in an argument, querying conclusions, or pointing to alternative interpretations. Most of us have had the experience of appreciating the careful work of a referee. Referees can also block the publication of a manuscript by identifying serious errors or omissions. A very few journals, such as the *Journal of Geophysical Research*, identify some reviewers at the end of each paper when it finally appears, but the majority of journals follow the traditional practice of secrecy. It should be noted, though, that some subspecialities operate within sufficiently small professional circles that an author can have a pretty good idea who will critique his or her manuscripts; still, the illusion of anonymity is a useful professional lubricant.

The process as I have described it has proved to be an effective way of exerting a control on the quality of the printed material.[2] Some papers sail quickly through the reviewing; others may be returned to their authors with suggestions for changes. Some manuscripts are rejected: They might be better suited to a different journal, the work does not meet the editorial standards, or the authors are unwilling to make the changes that the referees have proposed. Sometimes an editor must adjudicate between referee disagreements or deal with an irate author whose mangled prose prevented reviewers from understanding his ideas. Not all editorial and referee decisions are correct even though beyond negotiation (after a time). Most scientists have had the experience of having to shop around to find another journal whose editors were less picky or its reviewers more understanding or (as some would assert) whose standards were lower. Because of the multiplicity of journals and normal human variations, many manuscripts can find a good home in this way.

We should be clear just what a refereed publication does and does not imply. It does *not* imply total freedom from error. Even the most conscientious scientists, referees, and editors can be misled, fail to notice some subtle flaws. We consider a paper published in a reputable journal as probably free of significant errors and as having content that warrants our serious attention. Later work may fail to confirm some findings or may show where errors were made or may prove the paper's conclusions to be wrong. This does not mean, however, that the original paper was without merit, for the research it stimulated might lead to important improvements in theory or experimental methods. For example, in 1969 Joe Weber of the University of Maryland reported the detection of gravitational radiation. From Einstein's General Theory of Relativity, it had been calculated that massive objects that are rapidly accelerated would radiate gravitational energy, much as an accelerated electron radiates electromagnetic energy. Weber had explored the theory and then designed and built the first gravitational wave detectors, one of which is now on display in the Smithsonian in Washington. His published results were surprising, as they implied a much greater intensity of radiated gravitational energy than other scientists had expected. Even now, after more than twenty years, there has still been no confirmation of Weber's findings, no direct detection of gravitational radiation. Einstein's theory has by now been confirmed with great precision, however, by very different measurements, through observations of the way in which a special pulsar is behaving. (Pulsars are neutron stars, remnants of the explosive evolution of some stars many times more massive than our Sun.) Weber's papers appeared in the *Physical Review,* a prestigious journal, so what (if anything) went awry? It now seems as though a combination of experimental and statistical artifacts only made it seem as though gravitational radiation had been detected. Newer and far more sensitive detecting systems have yet to see anything. Though Weber has argued strenuously on behalf of his findings, he has not persuaded the majority of those able to judge. But the design of experimental systems to detect gravitational radiation was certainly stimulated by Weber's pioneering attack.

A second example of a reported observation that has not been confirmed comes from the 1987 supernova explosion. There was great excitement in February of that year when scientists saw the explosive brightening of an otherwise unremarkable star in the Large Magellanic Cloud, one of our galaxy's closest neighbors. This was the closest supernova to be seen since a 1604 event well documented by Johannes Kepler and Galileo, among others. What is especially important about the 1987 event was that for the first time a battery of modern astronomical techniques could be used on a relatively nearby supernova. As a result, both the quantity and the quality of the information are unique. The evolution of this remarkable object continues to be followed, providing us with much greater insight than we have so far had. So it was a matter of considerable importance when in March 1989 some radio astronomers reported detecting a pulsar at the site of SN 1987A. Unfortunately, no other astronomers could confirm this detection. Careful checking has since shown that radio interference from a television system was responsible for the spurious signal that mimicked the pulsar.

A third example of unconfirmed reports involves the alleged discovery of a monopole in 1975. Monopoles play an important role in the bestiary of fundamental particles. An ordinary magnet always behaves as though it has two poles, one north and the other south. We use this property when making a simple compass. A single pole (monopole) has never been seen, but this is a theoretical puzzle, for there is no reason for it not to exist. The report of discovery quite naturally provoked considerable interest, and the discoverer conducted seminars and was interviewed on the BBC, giving imaginative forecasts of the ways in which monopoles could be harnessed. Unfortunately, there were several very mundane alternative explanations for what he had seen, as several of us promptly pointed out through the *Physical Review Letters.*

What these several episodes point to is that highly competent scientists can and do err or be misled on occasion, but the professional community will usually spot the weak points or omissions very quickly, providing explanations well within the prevailing paradigms. No formal retraction is needed. Sometimes the error can be traced to hasty or careless work; sometimes we never find the cause.

I cite these examples to show that the professional literature does indeed contain reports that later turn out to be incorrect or are never confirmed. Although conscientious refereeing is not clairvoyant, it is generally effective. In the end the journal's reputation depends on the general level of papers it publishes and not on the occasional erroneous paper. In science there is no need to compile statistics of errors as in baseball; work at the forefront of science carries risks, and it need not be dishonorable to err. This is very different from deliberate fraud (which does occur and is the subject of Chapter 10).

The variety of scientific journals is enormous. There are, of course, journals devoted to chemistry, geology, archaeology, and so on, and each subject has an apparently endless series of subdivisions, each with its own publications. The *Journal of Geological Research,* for example, appears in four sections each month: "Space Physics," "Solid Earth and Planets," "Oceans," and "Atmospheres" (somewhat reminiscent of the ancient elements of air, earth, water, and fire). Within a single subject, there are journals devoted to theory, experiment, computing. Cutting across disciplinary lines are a very few general journals such as the weeklies *Nature,* which has been appearing for over 125 years, and *Science,* the official publication of the American Association for the Advancement of Science (AAAS). Both of these accept short original manuscripts in any area of science, but a significant part of each issue is given over to the reporting of general scientific news—short and nontechnical news items, notes of national and international aspects of science policy, and book reviews. Also unusual in these two journals is the regular use of editorials to express opinions and correspondence columns in which readers regularly disagree with the editor, the contributors, and one another. Those sections open to original reports are fully refereed; the general sections provide us a valuable window onto areas outside our own specialities. This evangelical function is the sole aim of the yet more general *Scientific American* and *Science News*, which do not carry any refereed reports of original work.

Much of the nature of the diversity of the scientific literature is apparent to any nonspecialist. Open any publication at random and you will find scientific jargon, a patois of abbreviation and convenience. Opaque at first sight, such language is not truly secret but is only gradually absorbed. Although the language may include mathematics as well as specialized terminology, it is open to all who make the effort to learn it.

What is *not* obvious to outsiders is that in each field the journals are informally ranked, by consensus of the professionals, in a hierarchy of prestige. In addition different journals meet different needs. The *American Journal of Physics* and the *Physical Review* both appear under the umbrella of the American Institute of Physics but differ vastly. Published by the American Physical Society, the *Physical Review* and its associated *Physical Review Letters*, on the one hand, are among the physics world's major journals. Their printed pages creep like an inexorable tide along our shelves, consuming about 10 feet each year, to the dismay of our librarians. One after another, major discoveries are reported in their pages. The *American Journal of Physics*, on the other hand, is published by the American Association of Physics Teachers, and it provides a valuable repository for the publication of physics pedagogy, apparatus notes, and reviews of physics books—but not new physics. We agree (never formally) that many of the journals that crowd our shelves and strain our budgets are helpful but not prestigious. We watch them and refer to their contents on many technical details, but we do not expect to find major discoveries announced in their pages. Like much of normal science, they are useful but unexciting, necessary but not glamorous. But some journals seem to have no obvious use apart from providing an outlet for pedestrian work that will never be cited except by its authors, who need to impress their deans and faculty promotion committees.

Some journals are published by learned societies, others commercially. Most major scientific societies publish their own journals. The *Proceedings of the Royal Society of London*, started in the seventeenth century, is the oldest journal still being published. Section A caters to the physical sciences, Section B to the biological. Another long-established journal is the *Philosophical Magazine* from the respected publishers Taylor and Francis (for so many years in the Red Lion Court in London); mammoth international publishing houses such as Springer Verlag and North Holland produce a wide range of publications including research monographs. Elsevier, publisher of Galileo's *Two New Sciences* in 1638, is still active. In all, thousands of journals pour out.

How do we keep track of these publications so that we learn of discoveries, improvements in techniques, advances in fields neighboring our own? For the most part, we confine our regular attention to a few journals that we know are closest to our interests and most likely to have what we want to know. We cannot read everything, and we take the calculated risk of ignoring the many minor journals that our professional gossip and our own experience tell us will be of little interest. But in order to cast a wider net, we make use of abstracting services. Almost all scientific papers start with a short summary or abstract outlining the main points. These abstracts are themselves collated into journals such as *Biological Abstracts, Physics Abstracts*, and *Chemical Abstracts*, which appear one, two, and four times each month, respectively, with comprehensive indexing and cross-listing. We go to annual summaries and five-year summaries when we want to check that we have not missed anything of importance. Through the abstracts we can easily locate original papers in any field or topic. Given the extent of the publishing enterprise (*Biological Abstracts* listed one-quarter million abstracts in 1989, *Chemical Abstracts* twice as many), the cross-referenced abstracts provide the best protection we have against being uninformed.

By now, you can probably understand why we want to guard the integrity of our printed pages. We simply could not afford the time and energy to locate nuggets in the mountain of print that we would encounter if the scientific journals had the editorial standards of the publications we find at our supermarket checkout counters. We place a very large responsibility on the editors and referees of our journals so that each piece that appears in print has a good chance of being correct, even if not very important. And still there is a huge volume of publications.

Professional societies do more to promote their sciences than publish their journals. With their membership usually restricted to people with appropriate credentials, either by training or experience, the societies often lobby in Washington on behalf of research funding, testify for or against proposed legislation, and work with other groups to promote science education or provide information to the media. They are linked with similar disciplinary groups in other countries in organizations such as the International Union of Geodesy and Geophysics and the International Union of Pure and Applied Chemistry, all in turn under the umbrella of the International Council of Scientific Unions.

Probably the most important and visible activity of the professional societies, in addition to their publications, is their sponsorship of conferences. Here a scientist's presentation can take on a dimension not possible in hard print. Some scientists are masters of the spoken word; others as persuasive as used-car salesmen. The majority are simply boringly efficient. Conference presentations fall into a few categories. Most meetings include a series of invited specialists who are asked either to report on their own work or to survey a topic and are given the luxury of about thirty minutes for their presentations. Contributed papers are not individually solicited but may be submitted by any member of the society; each is usually allotted between five and ten minutes to speak—barely enough time to advertise one's wares before the buzzer sounds.

There has been a steady growth in the number of scientists attending meetings. When the 1904 International Exposition was being planned for St. Louis, a team spent a summer traveling around Europe, personally delivering invitations from President Theodore Roosevelt and reporting back on the great difficulty they had in persuading busy scientists to stop their work and come to St. Louis to deliver papers. Only a small number of scientists seem to have attended those sessions. A famous photograph of the 1927 Solvay conference shows the all-star cast of fewer than thirty, including Einstein, Marie Curie, Ernest Rutherford, Max Planck, and Arthur Compton. Today things are different. There are a few very limited invitational conferences, such as those held by the Pontifical Academy in the Vatican, but there are also larger gatherings that draw hundreds or even thousands of active workers. A few major scientists are even said to spend more time on airplanes than at their home institutions. To accommodate the vast increase in attendance, many societies have adopted the strategy of holding parallel sessions and also having "poster sessions" during which contributors display a poster that outlines the main content of their "papers" and then hang around to distribute printed versions and answer questions. But for many of us, the most valuable part of a conference is the opportunity to meet and talk at length with our colleagues. By this means we become aware of work in progress long before a published paper appears or even before a conference presentation is made. Sometimes termed an invisible college, this informal sharing of ideas and results—sometimes in person, sometimes by letter, and now by e-mail—is an old tradition. The founders of the Royal Society met in just this informal way in Gresham College before the society was officially established by royal charter.

What I have described up to this point are the structure and operation of the scientific communication channels. I now turn to the ways in which the results of all this work become codified and then passed along to the next generation of scientists. We need to take a look at an often neglected component, the science textbook. Kuhn is probably alone in drawing attention to the importance of textbooks in shaping the scientists' attitudes to their craft. Why is this seemingly minor topic so important? Scientists talk to other scientists, contribute to and read journals, and go to conferences, but undergraduate students do none of this, though a few will become involved in research projects under the direction of a faculty member. These activities are well known to working scientists, yet virtually none of them are part of a scientist's education.

Science education is designed to produce graduates who either continue with a research education or enter the work force. Through the undergraduate years, the focus is very narrowly on the accumulation of factual knowledge, the content of science, and the acquisition of special skills. Introspection in the scientific method is almost unknown. Courses follow courses in increasing sophistication with (in some subjects) the expansion of the arsenal of mathematical methods to use in working a variety of standardized problems. Students learn laws and formulas in order to apply them and to recognize which law fits which problem. In more descriptive subjects there is a relentless parade of current knowledge, of established facts and their syntheses into current paradigms. Textbooks are constructed accordingly, with their endless progression of successful methods and accepted knowledge. Little or no time is spent in the undergraduate years in discussing the lessons to be drawn from scientific error or even describing the false starts, near misses, or theories that were useful for a time. There is no sense of continuity, of tradition. Instead, a pantheon without foundations is displayed, apparently occupied almost exclusively by the scientists who have laws or effects named after them.

To a degree, we have to sympathize with this strategy. We expect fledgling scientists to know much more than *we* did at that stage. Most instructors are unhappy about giving up valuable class time to the history of science or a discussion of its methods or even simply to ruminate. Our textbooks instruct us in the successes of science, but almost none has a section on the failures, the theories that were useful but later shown to be wrong. Not only is the resulting textbook picture incomplete, it is misleading. The mammoth 800–1,000-page encyclopedias that pass as freshman physics texts do occasionally have a few pages for history or editorial comments by famous physicists, but experience shows that these sections do not receive the same attention as those that carry material that might turn up in an exam. The net result is that after four years our students graduate with considerable technical proficiency but no historical roots and no component of skepticism. During graduate study and apprenticeship with a faculty member, students begin to see the other side of science, but their graduate texts still are directed to skills and knowledge. As I know from my own experience running seminars on the topics that I cover in this book, most graduate students have never examined our assumptions or looked at the organization of our science.

In this chapter and the preceding one, I have addressed a number of topics that might seem to be only slightly related, but I wanted to provide background for the important question that I discuss in the following chapters: What happens when unconventional ideas are put forward or findings are rejected and accepted theories are challenged? Our reactions as scientists are governed largely by our education and our experience. Much of our education does not prepare us for the inconsistencies we will encounter later in our careers. We welcome discovery yet tend to be conservative. Such attitudes are not really in conflict. Where we have a theory or paradigm that has served us well, we will not jettison it lightly.

Someone who proposes a radical change carries the initial burden of proof, the responsibility to persuade a referee and editor that there is sufficient merit in the proposal to warrant publication. Even publication in a respected journal is not a guarantee of correctness, but it will probably stimulate prompt attention. For example, in 1989 two Japanese physicists reported in the *Physical Review Letters* that the weight of a high-speed gyroscope depended on its speed as well as its direction of rotation.[3] If correct, this is the closest that we have yet come to a demonstration of antigravity. Their paper survived refereeing; there were no obvious flaws, although there were doubts. After publication, in short order, several scientists attempted to repeat the experiment with greater sensitivity and much greater attention to eliminating instrumental effects. The Japanese physicists' findings had generated surprise not only at the type of effect reported but also at its size; it was far greater than the limits of detectability and large enough to have important implications, if correct. It now seems certain that the original experiment did not include sufficient precautions against systematic effects. What we see in this example is the way in which the burden of proof shifts: The journal editors and referees were satisfied that a plausible case had been made. With the wider interest brought by publication, it did not take long before the result was challenged and shown to be in error. The scientific method worked well.

You will have noticed that I have phrased my description of the discharge of the initial burden of proof in terms of publication through a regular professional journal. Scientists do not greet with respect every putative revolutionary science that is announced. At one time, of course, major scientific ideas first appeared in books—Galileo's *Two Chief World Systems,* Newton's *Principia,* and Darwin's *Origin of Species.* None of these monumental works received peer review, none had to survive editorial scrutiny by a professional scientist. Galileo had circumvented church censorship, but that was not comparable to today's refereeing. Those days have long gone. Scientists now view with great suspicion virtually any book in which remarkable claims are made for the first time, as they generally consider this a sign of unscientific eccentricity.

Isn't this unfair? Perhaps there *is* some independent thinker, not part of the scientific establishment, who prefers to publish outside the mainstream or has been rejected by editors and is now using the only remaining means of getting his ideas out. Perhaps—but the chance that such ideas are correct is remote, and the "science" that comes to us first through books, magazine articles, and talk shows is pseudoscience or worse.

But how can truly independent thinkers receive a fair hearing if they are denied access to the scientific literature and then treated as cranks when, somehow, they publish their ideas? Does not the system amount to an effective form of censorship? The answer to this apparently reasonable question is yes: There *is* a system of censorship, and a very deliberate one. As I have mentioned before, the approval of the scientific community must in the end be obtained. If a new idea cannot survive the glare of an editor and referees, it probably needs overhaul or worse. Publication via the extrascientific media will not gain scientists' recognition.

An interesting experiment, a creative alternative, was introduced by *Nature* in August 1990. Until then, *Nature* had published its contributions under the separate headings of "Letters," "Articles," and "Review Articles." It now added "Hypotheses," intended as an occasional vehicle for scientific papers that failed to win the full-throated approval of the referees but that were nevertheless judged to be of sufficient importance to command the interest and attention of readers. The first article published under this heading set out a cosmological alternative to the currently accepted big bang theory. Its authors were Halton Arp, Geoffrey Burbidge, Fred Hoyle, J. V. Narlikar, and N. C. Wickramasinghe, all of whom had long lists of reputable publications but were also known for their often iconoclastic views.

"We are ... glad," wrote the editors, "to publish this absorbing document, in the spirit of constructive 'consciousness-raising.' ... It is not just another half-baked idea. ... It is not expected that contributions under this rubric will be frequent."[4] We will have to wait to see what next emerges in this section and whether other journals will take similar steps. Will scientists be willing to have their pet ideas appear in this section? How unorthodox can an idea be and still be published? This is an inventive and perhaps even courageous move, and even if the experiment does not survive, we should still learn something from it.

There is no single authority or oracle to whom the community turns for a definitive ruling that will, like decisions of the Supreme Court, have the force of law in pronouncing on the validity of some results or theory. These can also never be established by a vote. Some years ago, there was a proposal that a 'science court' be established to adjudicate on matters of scientific validity. Partisan presentations would be made with cross-examination permitted. The scientific community showed little interest in this idea.

There are situations, however, where scientists are called upon to serve as expert witnesses, when scientific and technological issues become matters of politics or legal dispute and both sides hope scientists' testimony will bolster their cases. But just who is an "expert"? A recent unanimous decision handed down by the U.S. Supreme Court will have a great influence on the way in which expert testimony is admitted in court. At issue was a case in which the defendants argued that the only scientific evidence that should be admitted as authoritative was that which had already gone through peer review, as would be shown by publication in a professional journal. In its decision the Court took the position that the trial judge should exercise his or her own discretion as to what should be admitted, with the provision that robust cross-examination would be allowed. But that still leaves open the definition of *authority* in science. Who is an "authority"?

Perhaps the best known of all modern scientists, legendary in his brilliance and sanctity, was Albert Einstein. Einstein's scientific fame rests on the remarkable sequence of papers that he published in 1905, followed ten years later by writings that set out the foundation for the General Theory of Relativity. His papers on relativity and gravitation appeared during World War I. Arthur Eddington in England realized their importance. Just as importantly, he saw the possibility of testing a predicted bending of light during the solar eclipse due for May 1919. Though not terribly accurate, the confirmation of Einstein's predictions was dramatic. There was no rival theory to explain the observations. Einstein's fame was ensured, his name suddenly the symbol of unfathomable genius—and authority.

(When the theoretical physicist Wolfgang Pauli showed signs of similar early brilliance, he became known as Zweistein.)

Which scientists today are as well known? Carl Sagan and Stephen Jay Gould do not rank with Einstein but are known to a wide public in the United States. Sagan has a reputation as a competent scientist of wide range; Gould is recognized for his work in paleontology. Sagan's flair as host of the popular TV series *Cosmos* established his position, while Gould, an electrifying speaker, is best known for the success of his many outstanding science books directed to a literate but nonspecialist public. Slightly less visible is Linus Pauling, whose first Nobel Prize rewarded his discoveries in chemistry, his second his efforts to eliminate nuclear weapons. Neither of these qualifies him as an expert on the use of vitamin C, though he has long championed its use. William Shockley, Nobel Prize winner as coinventor of the transistor, was much better known in his later years for his views on the relation between race and intelligence. Hans Bethe and Edward Teller are both theoretical physicists of great brilliance. Bethe, with a research career spanning more than sixty years and a winner of the Nobel Prize, has a commanding stature, someone considered almost never wrong. Teller, in contrast, has been at the center of many controversies. He is acknowledged to have many good ideas but rather more that do not pan out. Nevertheless, his political influence has been notable.

Over the years Bethe and Teller have testified, on opposite sides, in the continuing debates on nuclear weapons. Gould was one of the scientists called as expert witnesses in an important case in Arkansas involving creation science. In all of these examples, we have scientists who first established themselves within their chosen fields. Some became more widely known through their popular writings, some by the quality of their thought and the forcefulness of their personalities. These are the scientists to whom we turn when we want expert opinion on major public issues, scientists whose opinions will be generally accepted as authoritative in the arena where science and public policy overlap. But for whom do these scientists speak? What qualifies them for such deference to their views? Why are they singled out for so much respect, even awe? Others may be better informed but do not carry the same public clout. The much larger number of scientists who testify daily on lesser issues as expert witnesses in our courts serve similar functions without having the same weight. In none of these situations can scientists do more than present their own views of the sum of scientific knowledge on the topic under review; they do not have any official mandate to speak for the scientific community, but the weight of their statements rests on the accumulated respect they have been accorded. As is so vividly shown in the nuclear weapons debates, experts can disagree—on the "facts," on their assessments of effects, and on predictions. Witness the similar spread of views on the Star Wars proposal for an antimissile defense system. It is clear that other (nonscientific) factors intrude into many judgments, and the public becomes confused when experts disagree. But scientific experts are still widely sought after.

We see the same situation, on a less public scale, within each science. When a major discovery breaks, there are authorities whose judgments we treat with great (but not necessarily final) respect. Their names are not household words, but they are known within their own circles. But although courts and legislatures may listen with deference to scientific experts and although scientists are often invited to express their opinions on all sorts of matters, no new idea is welcomed into scientific knowledge just because of some authoritative endorsement. The movement to consensus is more like the advance of volcanic lava flow: usually slow, impossible to divert, and often accompanied by heat and odors.

There is another side to science that I have not yet touched on, involving the interaction of science and scientists with the rest of society. It is clear that the discoveries of science have been adapted by society for good as well as for ill; the residual responsibility of scientists for the uses to which their discoveries are put is the subject of continuing debate. It is equally clear that society, through its financial and other support for science, expects returns and some voice in the selection of subjects of research. Most of this vast and important area is beyond the scope that I have chosen for this book.[5] The one comment I wish to make here is on the objectivity of scientists.

Critics of science often assert that scientists are not as objective as they claim to be, that their interpretations are not value-free. I think that this assertion should be considered in two parts. Most scientists are, I think, objective on internal matters, as regards the content of basic science. Attempts to force one's experiments into preferred directions cannot succeed for long, and the constant exposure of one's publications acts as an effective air freshener. Occasional dishonesty, as described in Chapter 10, is usually detected quite soon and has no long-term effect on the content of science. Personal bias is more likely to enter into the interpretation of the observations. As regards experimental design, the open scientific literature ensures that bias will usually be promptly noted, but there are many areas where the prejudices of the investigator can intrude. For example, the size of the brain cavity in skulls can be measured. The extrapolation from there to the construction of a theory of human intelligence was more than merely subjective: It was clearly pseudoscientific and distinctly racist. Einstein nicely described the problem: "Even scholars of audacious spirit and fine instinct can be obstructed in the interpretation of facts by philosophical prejudices."[6]

I do not wish to minimize the importance of this and other aspects of the mutual interaction of science and society. However, I do not believe that these issues and their debates need affect our judgment of what is science or pseudoscience within the natural sciences. As we move through the biological to the social and behavioral sciences, there is indeed a much greater opportunity for extrascientific preferences to intrude. This is certainly the case in the IQ controversy and the debates over the theory of evolution. It is hard for me to see how personal bias can have any parallel influence on the calculation of the energy levels of the hydrogen atom or the distances to quasars or the viability of the cold fusion claims, all of which are open to quantitative testing.

Plate tectonics does not rest on political theory, even though Russian geologists were slow to accept it; the effectiveness of the CAT scanner is not dependent on one's social values, though its availability and cost are; the errors in the science of the creation scientists can be identified by all except the creationists themselves. Society can and does influence which scientific research is funded, but the scientific conclusions that are drawn from the research, the judgments that some practices and results are scientific while others are pseudoscientific—those judgments are as value-free as we can reasonably expect, and only this is of immediate relevance here.

Let me now sum up. Science has a working system in which new ideas are considered and tested. Each of us does not have to test every new result. Through journals and the opinions of experts, we distribute the load of testing, and we generally accept the opinions of those who have the expertise and have taken the time to examine the proposal in detail. We are more likely to accept less controversial ideas; the more radical the idea, the greater our reservation. The system is built on trust, a belief that the opinions are thoughtful and honest, not swayed by personal or other bias. When this trust is violated, we have a very serious problem; in Chapter 10 I review some examples of fraud. To repeat, we have a system far more complex than the scientific method of popular impression; its successes are demonstrated daily. But though it works remarkably well, it is not perfect.

We are now ready to examine a number of instances of revolutionary claims, and we will see how this system works in practice and how science both protects itself and also copes with innovation. In subsequent chapters we will also see how the system copes with fringe science and pseudoscience.

Notes

1. *Social Studies of Science* (1984): 521.

2. Peer review has come in for some strong criticism from scientists who (perhaps correctly) feel that their manuscripts have not received fair reviews and from nonscientists whose view of science is often quaint. For one of these dissenting views, see Daryl E. Chubin and Edward J. Hackett, *Peerless Science* (Albany: State University of New York Press, 1990).

3. Hideo Hayasaka and Sakae Takeuchi, *Physical Review Letters* 63 (1989): 2701.

4. *Nature* 346 (1990): 780, 807.

5. Michael W. Friedlander, *The Conduct of Science* (Englewood Cliffs, N.J.: Prentice-Hall, 1972).

6. P. A. Schilpp, ed., *Albert Einstein: Philosopher-Scientist* (New York: Harper and Row, 1949), 49.

6

Respectable Maverick Ideas

Reports of experimental discoveries and new theories provide excitement in science. Our work would be very dull if science were no more than the search for yet one more decimal place, the confirmation once again of a theory long accepted. The mistaken view that physics had exhausted all possibilities was actually expressed at the end of the nineteenth century, but then came the deluge: radioactivity, relativity, quantum theory, and nuclear physics. Of course the additional decimals do have utility, but what adds kick to our scientific morning coffee is the headline announcing something truly new, totally unexpected. At various places along the way in this volume, I have described the receptions that greeted a few innovations or the claims that were being made. I want now to sketch (in less detail) several more cases from which I can extract the common factors so that we can see general features. It is sometimes alleged that scientists are not open to innovation. What I hope these examples will show is that in reality many of the theories that we have entertained are just as bizarre as those that we have rejected out of hand, but that they differ in some very important ways.

This approach can be divided into three broad but not totally separate categories. In the first the innovations emerge within the family, so to speak, and the ideas come from scientists whose credentials are acceptable, established by their publication records. But credentials are not transferable, and the picture is less clear when a scientist ventures into a field in which he is not known. This will generally bring out strong criticism from those already entrenched in that field, who can see all of the omissions, the signs of the amateur. The second category is occupied by scientists whose credentials have (until then) been untarnished but who produce such improbable-sounding proposals that they have crossed the line separating the revolutionary from the incredible. Nevertheless, some reputable scientists will take the time to look into these often surprising proposals. Cold fusion is a prime example in this class. Examples in the third category have the common distinction of being promptly labeled as nutty by the experts, though these ideas are often welcomed by people less well informed (or more open-minded, according to your point of view).

The first group, the subject of this chapter, is, broadly speaking, treated with critical respect by scientists, as demonstrated by the substantive debate—which may, however, be robustly pursued. Some of the claims may be accepted after further examination, some are rejected, and, as we will see, the jury is still out on some. In marked contrast, none of the third group has been validated, though some still have their defenders, sniping from their bunkers but adding nothing to our fund of scientific knowledge. This group is the stuff of genuine pseudoscience, and I give them a chapter (Chapter 8) to themselves. The second (middle group) is also interesting: respected scientists who have erred or been sloppy in their work and who often hold tenaciously to their ideas, even in the face of having been discredited. These, too, have their own chapter (Chapter 7).

Dinosaurs

A good place to start our survey of apparently wild but still good science is with the theory explaining the extinction of dinosaurs. Dinosaurs occupy a unique place in the public affection. They hold a special fascination for small children— you surely know some three- or four-year-old who can tell the difference between tyrannosaurus rex, brontosaurus, and all the others. (This awareness seems to have been confined to children in the United States, at least until the appearance of the movie *Jurassic Park*.) Whereas other popular icons are extensions of imagination, the dinosaurs were very real. And, as we know, they became extinct many millions of years ago. The reasons for their total demise are the subjects of scholarly study and speculation. The widely held view is that the dinosaurs were unable to adapt to climate changes that other species survived.

An elaboration of this general view was put forward in 1980 by a team led by the Nobel Prize–winning physicist Luis Alvarez, with his geologist son, Walter, and two chemists, Helen Michel and Frank Asaro.[1] In their examination of the chemical composition of geologic strata set down over the ages, they found an anomalous concentration of the element iridium at the boundary between the Cretaceous and Tertiary layers. This transition is often termed the K-T boundary and has been dated to 65 million years ago, just about the time that the dinosaurs and many other species vanished. The Alvarez team pointed to the iridium abundance also known to occur in a certain class of meteorites, and they proposed a scenario that tied these observations together. What the Alvarez team now suggested was that a minor planet had collided with the earth with such violence that it was destroyed, throwing up vast quantities of dust, from itself and the earth, into the atmosphere. So dense was the resulting worldwide dust envelope that less sunlight could reach the surface, thus affecting the climate and vegetation. The iridium represents the clue to the nature of the cosmic projectile, and the K-T boundary dates the event.[2]

Minor planets (sometimes called asteroids because of their starlike appearance far from the earth) are rocks, typically tens of miles across, small on a planetary scale, that orbit the Sun. Over 2,000 minor planets have been discovered. Fewer than 10 percent have diameters greater than about 60 miles, and most are less than half that size. There is one group, the Apollo objects, whose orbits cross that of the earth, making collisions possible. Scientists believe that the minor planets and meteorites have similar compositions and that both formed during the early years of the solar system. We know that the earth has suffered occasional major impacts, and not only in the distant past, as shown, for example, by Meteor Crater in Arizona (formed about 50,000 years ago) and the impact at Tonguska in Siberia in 1908.

This ingenious cross-disciplinary proposal has raised great interest, both scientific and public. It has been attacked, defended, and elaborated. It is persuasive but not yet established. The presence of the iridium at the K-T boundary has been repeatedly demonstrated worldwide, but there is considerable disagreement even on the suddenness of the species' extinction and certainly on the mechanism. Many paleontologists hold that the dinosaurs and other species did not vanish suddenly but gradually, starting before the K-T date.

The research results have appeared in prestigious journals; conferences have been devoted to this theory. The regular and accepted channels for scientific debate are being used. This is stimulating science, conducted by respected scientists who are challenging accepted ideas. During the past few years, scientists have entertained a competing hypothesis: that the dust that obscured the sunlight and produced major changes in climate came from volcanic activity over many years. According to this scenario, a greenhouse effect resulted from the carbon dioxide released in these volcanic eruptions, and the subsequent climate change led to the extinctions. The volcanism theory is not now widely favored. Although we do not yet have a definitive explanation for the extinctions, the impact theory seems reasonable, and science has been enriched by this debate.

There have been two interesting and unrelated uses made of this research. For some years, there has been discussion of a possible "nuclear winter," the hypothetical consequence of a massive exchange of nuclear weapons between the superpowers. Proponents have suggested that so much dirt would be thrown into the atmosphere both from the bomb blasts and from the resulting fires that there would be a global cooling because of the absorption of sunlight. They go on to suggest that the cooling could be severe enough to cause climatic changes and perhaps even trigger the onset of an ice age. Opponents have contested this extrapolation from current knowledge and believe the effects would be far less severe and certainly not environmentally threatening. By showing how climate change could produce major effects on the population of some species, the Alvarez idea at first seemed to give support to the nuclear winter proponents. Further study, via computer and climatic modeling, has not supported the extreme nuclear winter projections. It should be remarked, though, that the 1884 eruption of Krakatoa did lead to a measurable drop in temperature, so we know that this sort of effect can occur, at least on a small though detectable scale. The original nuclear winter calculations appear to be have been oversimplified, but the problem is related to the suggested global warming and continues to receive attention.

An attempted application of a different sort has come from supporters of Velikovsky who have tried to use this work to lend support for their (or his) version of catastrophism. If cosmic projectiles are acceptable to explain the extinction of dinosaurs, why should improbable catastrophes not also operate as Velikovsky's theory had suggested? His was science by innuendo, and not very persuasive at that. The dinosaur and nuclear winter ideas, in contrast, were much more firmly based and did not rest on carefully selected readings and misreadings from the literature.

Relativity

Probably no other scientific theory during the past century has inspired the same combination of scientific acceptance with public mystification and misunderstanding. More than any other theory, relativity continues to attract the attention of cranks resentful of its denial of common sense and dedicated to presenting their own theories. What is the problem?

There are two theories of relativity. The Special Theory, enunciated by Einstein in 1905, describes the ways in which we perceive objects when they are moving close to the speed of light. The General Theory, published about ten years later, deals with gravitation and the shape of space near very massive objects.

Two assumptions are basic to the Special Theory: first, that the speed of light is the same no matter how measured and, second, that the laws of physics should have the same formulation regardless of the circumstances of measurement, whether we are moving or not when we make our tests. These seemingly innocent assumptions lead to some surprising predictions. For example, we are accustomed to the simple addition of speeds. If I can throw a ball with a speed of 50 mph, it will go a certain distance. As we all know, if I run at 10 mph and then throw, the ball will go further because it now starts out at 60 mph. This simple addition works perfectly well for moderate speeds, the speeds that we are familiar with, even with high-speed planes and rockets. When we deal with speeds above about 10 percent of the speed of light, we begin to notice departures from the predictions of simple addition. In these and other circumstances, we need to use the Special Theory, which has been repeatedly and convincingly checked to great accuracy and under a wide variety of experimental conditions. Well, the tests have been convincing to physicists but clearly not to everyone, and the theory's predictions are certainly not what our experience or common sense would have anticipated.

From its inception, relativity required us to be much more careful than we previously were in defining and describing the measurement of time. The true nature of time remains a fascinating subject. Time is usually considered to possess some absolute quality, flowing, as Newton said, "equably without relation to anything external."[3] Not so, Einstein maintained: The extent of an interval of time does depend on the relative motion between the observer and the subject of that observation. A major conclusion drawn from the theory is that intervals of time do seem to depend on motion, of ourselves as well as of the object being measured. An interval of time that is only one-millionth of a second long when nothing is moving can appear to be a second or even longer if the clock zips by us with a speed that is within a part in a trillion of the speed of light. Particles moving close to the speed of light have been observed among the cosmic rays and also coming from the higher-energy versions of cyclotrons. At 10 percent of the speed of light, the effect amounts to half of 1 percent, and that is easily measurable. Even more bizarre effects can result. Under special circumstances the sequence of occurrences can appear to be reversed: When nothing is moving, if A happens first and is followed by B, we might instead see B first followed by A if we rush by at high enough speed. What is happening to "reality"?

This insult to our common sense has not passed without strenuous challenge. Many people remain perplexed by this inversion of their intuition, unconvinced by the theory's alleged validity. I wonder how many of us who habitually use the equations and results of special relativity ever pause to reflect on how calloused we have become. We use the equations because they give us the right answers. Do we pause to ask what we mean by the "right" answers? We mean that high-speed nuclear particles turn up in the places where we expect them, and when they bounce off other high-speed particles, we can predict the outcome of the collisions.

One of the unexpected findings in Einstein's first paper was that mass and energy can be converted from one into the other, with the exchange described by the famous $E = mc^2$ equation, where c represents the speed of light. Physicists have used this equation in numberless nuclear experiments and get theoretical predictions that agree with the measurements. The answers are "right" to remarkable accuracy. That tells us that we know how to make the calculations, that the theory is "correct," but it does not mean that we "understand" it. Our consciences are not troubled by concern for those people who find relativity an assault on their commonsense views of the world.

For many years, relativity was not exactly welcomed by scientists. In his presidential address to the American Physical Society in 1911, William F. Magie, professor of physics at Princeton University, surely expressed the feelings of many when he called the theory and its basic assumptions

a great and serious retrograde step in the development of speculative physics. ... How will [the supporters of the theory] explain the plain facts of optics? ... They are asking us to abandon what has furnished a sound basis for the interpretation of phenomena and for constructive work in order to preserve the universality of a metaphysical postulate. ... To be really serviceable [a theory] must be intelligible to everybody, to the common man as well as to the trained scholar.[4]

In 1912 Louis Trenchard More, another physics professor, expressed his dissatisfaction in an article in the *Nation,* adding his concerns on relativity to those on Planck's still new quantum theory:

> Both Professor Einstein's theory of Relativity and Professor Planck's theory of Quanta are proclaimed somewhat noisily to be the greatest revolutions in scientific method since the time of Newton. That they are revolutionary there can be no doubt, in so far as they substitute mathematical symbols as the basis of science and deny that any concrete experience underlies these symbols. ... Undoubtedly the German mind is prone to carry a theory to its logical conclusion, even if it leads into unfathomable depths. On the other hand, Anglo-Saxons are apt to demand a practical result, even at the expense of logic.[5]

(Similarly disapproving comments greeted the paintings of the surrealists, and the comments of music critics after they had heard many now famous compositions for the first time are certainly entertaining. But science was supposed to be different, more objective.) Less expert views came from the *New York Times* in 1928: "Understanding the new physics is like the physical universe itself. We cannot grasp it by sequential thinking. We can only hope for dim enlightenment."[6] This sense of bewilderment continues even today, as I know from my own teaching experiences. The puzzlement was once well summed up by an undergraduate instructor of mine, who, when pressed for an answer as to which time interval was longest, told us that "it depends on who looked first."

I cannot close this section without pointing out that the good old-fashioned mechanics of Newton are still precise enough for everyday affairs, even the navigation of spacecraft. It is when we come to the high-speed particles that arrive as cosmic rays or emerge from our accelerators that we have no alternative to the use of relativity. There are in addition some low-speed phenomena where theoretical insight demands relativity, but many scientists never find a need to call on Einstein for help. In the example of relativity, the validation requirements have been met with great precision, but compliance with common sense cannot be a deciding factor in the evaluation of a theory. Science has absorbed Einstein's ideas.

Fifth Force

Everyone knows the phenomenon of gravity. What goes up must come down. In his 1687 *Principia,* Newton formulated the scientific description that we still use for calculating the everyday effects of gravity. With further experimenting, scientists came to recognize that other kinds of force also exist in nature, and by now we have been able to show that there are four basic forces: gravitational, electrical, and two kinds of nuclear force. Each dominates under certain circumstances. In addition to keeping us on the surface of the earth, gravity controls the motion of large objects often separated by vast distances, such as the planets orbiting the sun. The electrical force is responsible for holding electrons to each atomic nucleus and for binding atoms into molecules. At much smaller distances the strong nuclear force controls the stability of the nuclei while another but weaker nuclear force governs radioactivity.

When we come across some new phenomenon, we first calculate the effects of the four known forces. Only if we find that these are insufficient to account for our new results do we consider the possibility of a new (fifth) force for the unexplained remainder. So far, when we find what appear to be additional types of force, we have always been able to show that they can be reduced to something more familiar. Magnetism, for instance, is not a separate phenomenon but we understand it to be basically electrical. The force we have to exert to pull a spring or a rubber band is also electrical, as we are pulling against the force between atoms.

Not content to leave well enough alone, some scientists are working in two ways on an understanding of forces. Some are trying to formulate sweeping theories that will combine the descriptions of all of the force in nature. Progress has been made, but the inclusion of gravity in this grand unification has yet to be achieved. Other scientists are on the alert for experimental evidence of a fifth force.

A flurry of excitement was triggered in 1986 by Ephraim Fischach and his colleagues at Purdue University.[7] They had reanalyzed the data of a classic gravitational experiment carried out by Baron Roland Eötvos in Budapest in the early years of the twentieth century. The Purdue group claimed to have found evidence for departures from what would be expected if gravity alone were at work. In their paper in the *Physical Review Letters,* they suggested that Eötvos's results could be explained if there was a very small difference in the gravitational forces on different kinds of atoms. Eötvos had carried out tests on wood and a variety of metals including copper and alloys of aluminum and magnesium. The absence of any difference in gravitational attraction on different materials had supposedly been demonstrated by Galileo in his famous but apocryphal experiment in Pisa, when he found that the rate of fall did not depend on the size or nature of what was falling. The universality of this effect is deeply embedded in Einstein's theory and had been the subject of some very sensitive experiments more recently. The Purdue analysis therefore came as a great surprise.

The response, as might be expected, was twofold. Theorists found an error in the Purdue calculations, and there is general agreement that it is now, at this late date, impossible to be sufficiently sure of all aspects of Eötvos's experimental conditions to be able to make the calculations to the accuracy needed for this test of gravity. The other approach has been experimental. The Purdue publication stimulated the design of much more sensitive experimental systems. To examine the effects of gravity, experimenters have taken their apparatus to the top of TV towers and down in mines. To look for a small possible effect of a new force, one must first know to even greater accuracy the effect of "well-known" gravity. I qualify this as "well known" because although the basic formula is well established, its actual use requires us to know the masses of all the surroundings that exert their gravitational attractive force on our experimental hardware. This is not as simple as it appears. The density of the earth's crust varies from place to place and also with depth. Mapping of the density variations has not been carried out to the accuracy needed. There have by now been many gravitational checks with an impressive variety of approaches, and by the early 1990s the consensus was that there is no convincing evidence for the fifth force.[8] This means that nothing has been detected at current levels of experimental sensitivity. Some scientists continue to pursue even greater sensitivity, for the basic question is still important, and it would be useful to place firmer limits on the magnitude of any additional force.

Cosmology

The origin and evolution of our universe have been the subjects of speculation for as long as we have recorded history. Theorizing used to be easy, as cosmologists could conveniently construct their theories without the possible embarrassment of data or the prospect of being tested. This has changed through advances in astronomy that allow the distances to stars and galaxies to be measured. At the same time there have been developments both of instruments to measure radiation from those distant regions as well as the theories to interpret their spectra. It was just such analysis of light from distant galaxies, by Edwin Hubble in the 1920s, that showed that the universe is expanding. Almost all galaxies are moving away from us at speeds proportional to their distances. (A few of our closest galactic neighbors are moving toward us, but this does not alter the overall picture.) Thus far, there is general agreement. Where opinions differed was in the explanation for this expansion. By 1950 there were two main cosmological theories, popularly known as steady state and big bang.

In the big bang picture, all of the matter in the universe was once confined within a very tiny volume at very high temperature. This minute region exploded, blowing everything out. The expansion that we detect today, about 15 billion years later, is the result of that primordial explosion. The alternative theory, steady state, accepts the reality of the expansion but postulates that expansion has always occurred and will always continue. Further, according to the steady state theory, the average number of galaxies in any region of the sky has not changed over the ages. But because of the expansion, the number of galaxies in any given region will decrease unless something happens to prevent this. To compensate for the expansion, new atoms are assumed to appear spontaneously in otherwise empty space. The newly created atoms will gradually combine, ultimately to form new stars and galaxies to replace those that have been carried outward by the universal expansion. In this way the average number of galaxies in any given region is maintained at a steady level.

As we catch our breath in pondering these grandiose cosmological scenarios, where do we begin to look for tests of the theories? So far, we know of only one universe, so we cannot exactly carry out a survey to look for general features common to all universes. We cannot construct another universe except in our imagination, and some theorists are doing just that, with computers. We now demand more from cosmological theories than simply plausibility or mathematical elegance. We require that they yield predictions that, in principle, are open to testing.

The theoretical basis of both the steady state and big bang theories came from Einstein's equations, but they differed in their assumptions and predictions. Cosmological theory has to conform to the present universe that we see and also describe the universe as it was in the past and will be in the future. We cannot wait long enough to check out the theories' predictions for the future, but we can investigate the past, in a way unique to astronomy. When we make our astronomical observations, we are actually looking back into the past. Sunlight that we are now enjoying left the sun eight minutes ago, but light from the bright stars has been traveling for thousands of years or more. Light from galaxies that are rushing away from us left them millions or billions of years ago. Here we do have a way to test our cosmological theories, for they disagree in their extrapolation into the past. Especially in the remote past, the big bang universe must have been very different from the steady state version.

We know that individual stars evolve. As they age, they burn the nuclear fuel in their cores; their luminosities (intrinsic brightness) must change. Galaxies, assemblies of stars, will therefore also change. Accordingly, as we peer into the past, the two theories predict differing numbers of galaxies with differing radiative powers at different distances in the past. This probing was carried out by optical and radio astronomers with inconclusive results. But the theory and observations were not well advanced in the early 1950s, and it was not possible to favor one or the other of the two theories.

The two theories diverged completely with regard to some other tests. For example, if the big bang theory was correct, the universe had started from a very dense and very hot stage, its temperature around 1 trillion degrees. As we know from the behavior of gases, expansion is accompanied by cooling unless extra energy is added. Theorists calculated that the universal temperature today should be far below freezing, probably below $-250°C$. But no matter how low the temperature, radiation will be present, and measurement of this radiation can allow the temperature to be calculated. An attempt to detect this residual radiation was made soon after the end of World War II, but without success.

There matters rested until 1965. In that year Arno Penzias and Robert Wilson of Bell Laboratories accidentally discovered this cosmic radiation, the glowing remnant from the big bang. They were puzzled by a small signal that their ultrasensitive radio antenna was detecting, although they thought that they had taken every precaution to eliminate unwanted interference. The temperature of the radiation that Penzias and Wilson detected was $-270°C$, only three degrees above absolute zero, the lowest temperature theoretically possible. Robert Dicke and his colleagues at Princeton were quick to identify the source of the Bell Labs' signal with the big bang theory, and they and others have abundantly verified the observations. The most recent and striking measurements have come from the COBE satellite NASA launched in 1989.

With this discovery acceptance of the big bang theory was strengthened, and it is now the currently accepted cosmological theory. Many other calculations have been made of other effects to be expected if the big bang scenario is correct, and the agreement has been good. One might think that the steady state theory has been eliminated, but as so often happens, a few scientists are not yet willing to abandon it. From time to time dissenting views appear, one of the most recent being the paper that was used in the first "Hypotheses" section in *Nature* (see Chapter 5).

What we have seen here is just how radical a theory scientists have been willing to entertain. Spontaneous creation of matter, as needed for the steady state theory, has not yet been detected, but then the suggested rate was only one atom in about 1 cubic foot in every 20 billion years. If astrophysical theory had been as well developed in 1950 as it is now, the steady state would probably not have been considered as a competitor; in those more innocent days, though, it held its place for a while. Whether the big bang will withstand further criticism is not clear. Although it has been modified to agree with more recent observations, its essential features remain. There can, however, be no doubt about the boldness of the cosmological theorizing, which at times seems to be only tenuously connected to reality.

Quantum Ideas

For a final example of a seemingly counterintuitive theory that has been absorbed into science, I turn to quantum mechanics—but only briefly because of its complexity. As with relativity, its more detailed workings are shielded from general scrutiny by its mathematical language. Also as with relativity, its content has often been misunderstood and then contorted and resisted. And finally, it, too, has been so successful in its incorporation into physics and its applications in chemistry that it faces no current credible challenges. This does not mean that it is a perfect theory beyond criticism or improvement. There is much research devoted to deepening our understanding of this complex theory, and there is learned disagreement over the implications of some of its results. Some current theoretical research attempts to combine quantum mechanics with gravity and the theories of other forces. But because of its complexity, most of this research passes unnoticed outside some well-confined theoretical circles.

To describe a theory as being counterintuitive or in disagreement with common sense introduces historical prejudices that frequently have little or no relation to a scientific understanding. Common sense uses our everyday experiences as the standard for judgment. The simple description of these experiences may be correct, but all too often our common sense is not a good guide as we seek to interpret those experiences. That is where the nonexpert may fail, for the theories that are successful can seem to defy common sense.

This has been a repeated story in physics with the study of mechanics, the science of motion. We all see things moving, but as repeated surveys have shown, popular understanding of these motions is often still primitive, and those of us who teach introductory physics have ample demonstration of this. The Newtonian description, now over 300 years old, has been absorbed by depressingly few people, understood by far fewer. Common sense can mislead us, as it did people from Aristotle to Galileo, and it is certainly not a reliable guide from which to attack relativity. We find a similar situation with quantum mechanics. When the equations of quantum mechanics describe an electron in an atom, for example, we do not get an exact prediction of where the electron is. What we do get is a statistical prediction: We are told how likely it is for the electron to be at this location or have this energy or at that location with a different energy or yet again somewhere else. The language is that of uncertainty, and indeed one of the most important results of quantum theory is Werner Heisenberg's uncertainty principle. This defines the maximum precision with which certain related quantities can be measured simultaneously. This imprecision is a serious problem at the atomic level but undetectable in our everyday affairs.

Quantum mechanics describes the emission and absorption of radiation by atoms and the ways in which nuclei behave when they collide. In one particularly famous experiment, the quantum mechanical description is formulated in such a way that it seems as though two electrons arriving at the same place sometimes cancel one another out and at other times add together. It is as though sometimes $1 + 1 = 2$ but at other times $1 + 1 = 0$. How do we deal with such a bizarre theory? In practice and in conduct of our day-to-day use of quantum theory, we close our eyes, take a deep breath, and follow the calculations—and arrive at the correct answer. We cannot tell where any particular electron will go in this sort of experiment, but we can, as accurately as needed, specify how many electrons will arrive at any particular place. Quantum theory really does work, notwithstanding our commonsense reservations. We can predict the likelihood that an electron will arrive at a given place, though this is less than certainty.

The theory has been expanded from its initial form as a theory of radiation (introduced by Planck in 1900). As its scope was enlarged, its name changed to quan-

tum mechanics to quantum electrodynamics and then to quantum chromo-
dynamics, and the end is not yet in sight. Disagreement still centers on the
interpretation or meaning of certain results—are they deep or trivial? It is,
though, a theory with a remarkable range of successes. Planck, who started all of
this, was well aware of the difficulty in setting aside one's common sense: He
commented that one could never really convert those who saw the quantum ideas
being introduced. As a new generation of scientists comes onto the scene, they
will take the new theory for granted, and the older generations will gradually die
out.

Conclusion

The ideas reviewed in this chapter have been far more revolutionary, far more
radical in their departures from their predecessors than any of the pseudoscientific "theories," yet they were examined with care even if initially with perplexity.
What are the ingredients of the theories just described that qualify them as respectable science, separating them from the quaint and the eccentric? There is no
single and simple litmus test. All of these theories have survived peer review to
appear in major professional journals. All have suggested tests. Relativity, cosmology, and quantum theory have prospered, though all three are still subject to review. Extensive experimenting has only strengthened our confidence in them.

There is an additional reason for their initial acceptance, at least to the extent of
meriting tests, and this comes from the scientists who devised them. They have
shown a clear understanding of their fields; in contrast, a hallmark of the pseudoscientist is an ignorance that the experts can spot at once, though their pseudosci-
ence-speak will often appear learned to nonscientists.

Notes

1. *Science* 208 (1980): 1095.

2. There is a large literature on this subject. See, for example, the article by E. William
Glen, *American Scientist* 78 (1990): 354, and David M. Raup, *The Nemesis Affair* (New York:
Norton, 1986).

3. *Principia: Mathematica,* ed. Florian Cajori (Berkeley: University of California Press,
1962), 6.

4. Quoted in L. Pearce Williams, ed., *Relativity Theory* (New York: John Wiley and Sons,
1968), 117.

5. Ibid., 121.

6. Ibid., 129.

7. *Physical Review Letters* 56 (1986): 3.

8. For good reviews, see L. M. Krauss, *Fifth Essence* (New York: Basic Books, 1991), and
Allan Franklin, *The Rise and Fall of the Fifth Force* (New York: American Institute of
Physics, 1993).

7

Walking on Water or
Skating on Thin Ice

A common thread that runs through the examples examined in Chapter 6 is that they have stimulated good research; science has been enriched by the underlying innovative ideas, even though not all have yet been proven correct. In contrast, the two "discoveries" that I discuss in this chapter probably will not leave their marks on the content of accepted scientific knowledge. Instead, both should be seen as cautionary lessons in the experimental artifacts that can arise and in the potential for deluding oneself unless one exercises great care.

Water is a central ingredient in the two scientific episodes I describe in this chapter. The first one deals with polywater, the hypothesized joining together of water molecules to explain some strange experimental results. The other case involves the claimed discovery of the ability of water to remember materials dissolved in it long after those materials have been removed. The cold fusion fiasco might also have been included in this chapter, as it, too, used a form of water as the agent for wondrous claims.

Beyond this damp connection these occupants of the penumbral zone surrounding mainline science share another feature, largely by exclusion. As we will see, they have not found a comfortable place among the more enduring components of science, but neither should they be classified with the clearly pseudoscientific and fraudulent cases that are the subjects of later chapters. They have come to share this chapter by default. Both involved reputable scientists pursuing fleeting targets. They showed carelessness, but that label, applied in hindsight, could not be easily awarded from the start. A level of disbelief was present from the beginning, and though this eventually won out, there was also an initial feeling that—just perhaps—something real was being seen.

The spectrum of scientific quality shades smoothly and imperceptibly from the genuine to the crank and the knowingly pseudoscientific. There is an adjacent spectrum of endurance. Some short-lived "discoveries" have proved to be valuable; others equally ephemeral have simply wasted our efforts. Drawing boundaries is to some extent a matter of organizational convenience, and some of the examples I have used could just as well move from one category to the next. Still, these two cases seem to group together and provide some interesting lessons.

Polywater

As this saga unfolds, you will surely notice close similarities to the development of the cold fusion case, though that one moved much more rapidly through its rise and decline.

Water is a remarkable substance. Its chemical composition of two atoms of hydrogen tied to one of oxygen (in chemical language H_2O) has long been known. In grade school we learn that the freezing point of water at 32°F (0°C) and its boiling point at 212°F (100°C) serve to standardize the temperature scales, and that ice, water, and steam are merely different manifestations of the same material. Without water, life as we know it could not exist. The properties of water, its behavior under different conditions of temperature and pressure, have been well examined, but water remains a center of important research, at the interface between physics and chemistry. Research topics include the behavior of the molecules of water when materials are dissolved in it, the complex structure of thin films of water on various surfaces, and the properties of very small droplets of water, as found in clouds. There is still a lot that we do not understand about water's behavior.

And so it was with considerable interest that knowledgeable scientists greeted Russian reports, in the mid-1960s, that some samples of water were displaying unusual properties under what appeared to be otherwise normal experimental conditions. "Anomalous water," as it was first called, was found to have a lower freezing point and higher boiling point than the well-established values. The density of this "anomalous water," its thermal properties, its behavior on surfaces all pointed to modification in the molecular structure of the water.

In their vapor form water molecules are free to bounce off one another and around their container like tennis balls. But in the liquid water and solid ice phases (forms), the molecules are packed close together, and adjacent molecules have a stronger and persistent influence on one another. In experiments where anomalous behavior was showing, it seemed that something was changing in the collective molecular behavior. In one experiment it showed up when silica gel containing water was heated. Nikolai Fedyakin, a scientist at the Technical Institute of Kostrama, discovered his anomalous water while examining water sealed inside capillaries, very narrow glass tubes. (The behavior of water in capillaries is important in plants and animals.) The behavior of this water on the glass surfaces did not follow known patterns. (You can see for yourself how water beads into droplets on a waxed surface or behaves differently when you add a few drops of detergent.) Boris Deryagin of the Institute of Physical Chemistry in Moscow used quartz tubes instead of glass and heated pure water in a container from which the air could be pumped out. The water in his sealed apparatus behaved in ways that Fedyakin had earlier seen.

Fedyakin's results were published in 1962 in the *Kolloid Zhurnal,* but the English translations appeared in the West only after about a year. (Most Western scientists cannot read Russian or Chinese, so for many years the Russian and Chinese journals have been translated into English, appearing from some months to a year after the original publication.) After the initial reports, most of the Russian papers on anomalous water came from the better-equipped Moscow laboratories rather than from Fedyakin, and Deryagin's papers started to appear in 1963. Fedyakin had not been previously known in the West, but Deryagin's international reputation was well established, as shown by *Scientific American*'s invitation to him to write an article on intermolecular forces as early as 1960. He was referred to as "a great physical chemist who has dominated theory and experiment in surface and colloid science for fifty years."[1] His work had been recognized with the Lomonosov Prize (roughly equivalent to the U.S. National Medal of Science) and the Order of the Red Banner of Labor.

The Russian work seems to have attracted little or no attention outside Russia for several years. Deryagin reported on his results at a 1963 conference, organized in Moscow by the International Union for Pure and Applied Chemistry, where some Western scientists were in attendance, but even in this forum the potential importance of the findings rang no bells. It was only at a conference in England in 1966 that Deryagin first had a large audience of Western scientists; once again, however, there were no major reactions. After that conference Deryagin visited several laboratories in Britain, and for the first time his strange findings began to prompt people to get involved.

Felix Franks, himself a chemist noted for his investigations on the properties of water and a participant in the subsequent campaign, has written a fascinating book in which he has chronicled the rise and fall of anomalous water.[2] His technical awareness and personal involvement have provided him with the basis for most insightful analysis and comments. Franks has shown how interest increased from this slow start and in spite of widespread and persistent skepticism. In 1965 only two papers on anomalous water appeared. This doubled by 1966, but until then all papers had been in Russian. When interest peaked in 1970, close to 100 papers were published in the United States and about another sixty elsewhere. By 1974 the numbers were down below five again. What had happened?

Surface experimenting is notoriously prone to contaminations, in the liquid and on the surfaces. Almost from the beginning, at least in the West, the cleanliness of the experiments was questioned. Deryagin reported on the precautions that had been taken to guard against impurities and the tests to verify cleanliness. By 1969, scientists outside Russia were publishing as many papers as those within. The mood in the West seems to have been shaped by skepticism, somewhat tempered by respect for Deryagin's reputation for quality work and by the growing

impression that a development of great importance was in the making. Nevertheless, doubts persisted; some experiments were inconsistent; some confirmed the
Russian findings.

As scientists sought to replicate the anomalous water results, explanations were
sought through theoretical speculation. There is often an irresistible temptation
to indulge in theorizing long before the results are securely established. Some of
this imaginative theorizing can be useful to experimenters as a guide to definitive
further experiments, but at times the theorists seem to have had both feet planted
firmly in midair. There is an allure to the thought of being first to come up with
the correct explanation, an impetus only slightly restrained by the greater prospect of being first to be demonstrably wrong. (Theorem: There are many more
wrong answers than right ones, and they are easier to find.) Anyway, it was suggested, if the water in these experiments displayed such strange properties, perhaps this was because the molecular structure had been changed, so that in
anomalous water the molecules reacted differently among themselves and with
the surfaces to which they came in contact.

Probably the turning point in public awareness came from a possibility raised
in a paper from a group at the U.S. National Bureau of Standards. They wondered
whether anomalous water was "polymeric water," or "polywater." They were suggesting the possibility that water molecules became linked into long chains or
polymers. Polymerization is the basis of the plastics industry. The properties of
these materials depend on the number and types of atoms in the molecules and
especially on the ways in which long chains of molecules are bonded together. The
art of producing new materials lies in the modification of those bonds and the
subtle manipulation of the proportions of different types of atoms. Now NBS
suggested that, under certain conditions, water might polymerize. The NBS
group reported spectroscopic evidence that pointed to the presence of inter-
atomic bonds not previously known, as well as the absence of a well-known bond
between oxygen and hydrogen atoms as seen in regular water. Their results also
showed no signs of contaminants that might have been dissolved out of the
quartz containers.

The polywater label was seized upon in popular reports and, as might have
been expected, ingenuity triumphed over caution in the extrapolation to fanciful
scenarios. The role of polywater in biological processes was considered. A paper
in *Nature* even raised the extreme possibility that polywater was potentially dangerous, for it could turn ordinary water into polywater and make it unusable by
plants and animals.[3] Some theorists produced a model for the molecular structure of polywater with water molecules arranged in clusters. Other theorists suggested a graphitelike structure with linking between flat layers of molecular chains
and loops.

In his November 1970 article in the *Scientific American*, Deryagin was still describing and defending the phenomenon of polywater, which he termed "Water
II." He devoted a long section to "skepticism concerning Water II" and concluded

his article with the opinion that "it is appropriate to end in the certainty that continued investigation of the properties of liquids will lead not only to universal acceptance of the reality of Water II but also to the discovery of new and even more surprising phenomena."[4]

Franks has labeled the stage reached by 1970 as "all aboard the bandwagon," but even at that time the tide was turning, as symposia were organized to focus on this perplexing liquid. More experimenters were finding evidence for contaminants. Some findings could not be replicated; some found traces of human sweat in the anomalous water. Along the way refereeing seems to have been capricious, as some papers were rejected yet similar papers accepted. The disinterested involvement of some of the journal referees has been questioned.

In his blow-by-blow account of the life of polywater, Franks identified the end of 1970 as the time when "the majority of experts were of the opinion that ... the eccentric properties of the substance known as polywater resulted from relatively high concentrations of foreign matter."[5] Although many questions remained unanswered, more and more scientists withdrew from polywater research. For many years, there has been no listing under *polywater* or *anomalous water* in *Physics Abstracts*.

Now, more than twenty years after the peak of interest in polywater, and especially with the availability of Franks's book, we can try to assess the scientific merits of polywater and the response of the scientists. The experiments of Fedyakin and Deryagin certainly displayed the characteristic signs of an anomaly, to use Kuhn's terminology. There was an anomaly but no crisis in sight as yet, nor did one develop. The reported behavior of the water, coming from reputable sources, definitely justified attempts to replicate and then to understand the reported phenomena. In these aspects polywater and cold fusion are similar, and both shared another important factor: the relatively inexpensive and apparently simple apparatus that was needed. No billion-dollar accelerator was needed, no expensive spacecraft with its attendant launch uncertainties.

In both cases the potential importance of the discoveries induced many scientists to join the chase. Where cold fusion and polywater differ is in the credentials of the early workers. Fedyakin was not well known, but Deryagin's reputation guaranteed that the anomalous water reports would sooner or later be taken seriously. In contrast, Fleischmann and Pons had creditable records in electrochemistry, but neither was known in nuclear physics, and from the start their results did not carry credibility with most physicists. The cold fusion work of the BYU and University of Arizona physicists, however, was taken more seriously, as they had been involved in fusion research for a number of years. Another difference between the cold fusion and polywater cases lies in the completeness of the documentation. The anomalous water papers, though delayed by the need for translation, did allow other scientists to try to reproduce those experiments. The secretiveness of Fleischmann and Pons turned replication into a sort of treasure hunt as physicists tried to guess the details of just what the pair had done. And at least the Russians did not have a lawyer who threatened legal action against scientists whose results contradicted theirs.

Both cases also show that subtle experimental effects can intrude and that scientists can in innocence be misled by all sorts of artifacts. In spite of awareness and precautions, contamination was indeed the problem with polywater. In the cold fusion experiments, the behavior of neutron and gamma-ray counters and the not-so-obvious problem of delicate thermal measurements misled many investigators. The requirement of replication is fine in principle, but it can be very difficult in practice, even if the equipment is available.

Whatever else these cases represented, they were not examples of pseudoscience. The discoverers of these effects were trained scientists who clearly believed that they had done careful work, and there was enough of the aroma of possible major discoveries to induce many scientists to divert their energies to attempts to replicate the findings.

The anomalous water case never displayed the circus atmosphere that the cold fusion discoverers generated from the start. But if you thought that the cold fusion case represented the extreme end of the spectrum, away from stolid and pedestrian science, just wait until you have read through the next section.

Footprints in the Water

"When to Believe the Unbelievable" was the heading to an editorial on the first page of the June 30, 1988, issue of *Nature*. The reference was to a relatively short (three-page) paper that carried the title "Human Basophil Degranulation Triggered by Very Dilute Antiserum Against IgE."[6] The title was seemingly innocuous, even opaque, to people in other research areas. There were thirteen authors from laboratories in France, Israel, Italy, and Canada, but the major author and the center of the subsequent controversy was Jacques Benveniste, of INSERM, a research institute at the Université-Paris-Sud. Remarkable as was this editorial commentary on a paper that, after all, the editor of *Nature* had agreed to publish, there was a second salvo directly following the paper. Under the title "Editorial Reservation," it ran:

> Readers of this article may share the incredulity of the many referees who have commented on several versions of it during the past several months. The essence of the result is that an aqueous solution of an antibody retains its ability to evoke a biological response even when diluted to such an extent that there is a negligible chance of there being a single molecule in any sample. There is no physical basis for such an activity. With the kind collaboration of Professor Benveniste, *Nature* has therefore arranged for independent investigators to observe repetitions of the experiments. A report of this investigation will appear shortly.[7]

Nature on rare occasions has commented editorially and critically on one of its papers, but I know of no other paper, in *Nature* or any other journal, that has been accorded this sort of accolade. None of the journals that I read carries the motto "Caveat emptor" on its masthead, yet the story gets even yet more bizarre. Read on.

The object of this editorial ambivalence was the reported biological effectiveness of aqueous solutions so dilute that no molecule of the presumed active ingredient was even present. It was being claimed that somehow the water molecules had been "imprinted" by their previously dissolved material. Later, and even after extreme dilution, the water was found to have retained (in some unknown way) a recollection of what it had once contained, and this recollection continued to possess the ability to stimulate changes in biological systems.

The *New Scientist* gave good descriptions of the experiments that had produced this strange result: "White blood cells called basophils, that form part of our immune system, carry certain antibodies (of the IgE type) on their surface. When these cells are exposed to antibodies directed against the IgE molecules, they release the chemical histamine stored in granules within the cells and become 'degranulated.' This changes the way the cells respond to certain chemical stains."[8] And, "In Benveniste's laboratory, researchers mix a measured volume of blood with one of the antibody dilutions and then stain the cells with a dye that turns intact basophils red. A researcher counts the red ones under a microscope, to calculate the proportion of reacting 'degranulated' cells."[9]

Not only did this degranulating power persist in the water long after all IgE molecules had been diluted away, but the degranulating power oscillated with the degree of dilution, increasing and decreasing with every tenfold dilution. Scientific credulity was being strained.

An added dimension of disbelief came about because the claimed medicinal power of extremely dilute materials is at the center of homeopathy, about a quarter of the physicians in France prescribe homeopathic remedies, and Benveniste's research laboratory received financial support from Laboratoires Boison, a major shareholder in a large manufacturer of homeopathic medicines, Laboratoires Homéopathique de France.[10]

Conventional medicine has not been sympathetic to homeopathy, and the publication of Benveniste's results in a journal as prestigious as *Nature* was in general thought to confer too much legitimacy to what was considered by many to be a hocus-pocus approach to the treatment of human ailments. Scientific reaction to the appearance of the Paris paper covered the predictable range. Opinions of various scientists were reported in the *New Scientist*: "I think it's rubbish, but it's one of the most interesting papers I've ever seen. It really does require some explanation"; "If it's right, there's some force working that we don't understand. Divine intervention is probably about as likely"; "It is a whole load of crap."[11]

Support, however, came from David Reilly of the Department of Medicine at the University of Glasgow's Royal Infirmary. Reilly had himself published a similar paper in the highly regarded British medical journal *Lancet* in 1986.[12] Though controversial, that paper had attracted nowhere near the same attention as did Benveniste's. Ian Munro, editor of *Lancet*, commenting on the publication of controversial papers, noted that "once I get to the point of deciding something is worth publishing, I light the blue touchpaper and await developments."[13]

When Benveniste submitted his manuscript to *Nature*, longtime editor John Maddox believed

> It is entirely possible for physicists to welcome the notion of a fifth force because it would be a novel happening which could nevertheless be accommodated within the accepted framework of science. Benveniste's observations, on the other hand, are startling not merely because they point to a novel phenomenon, but because they strike at the roots of two centuries of observation and rationalization of physical phenomena.[14]

Nature's referee reports had been highly critical, and as a condition of publication, Maddox required Benveniste to seek confirmation of his findings from other laboratories, hence the multiple laboratory and author listing in the finally submitted paper, which appeared after two years. Even then, Maddox was not satisfied, and, as he mentioned in his "Editorial Reservation," *Nature* "arranged for independent investigators" to review the experiments.

The very occurrence of that investigation as well as its actual conduct attracted the next lightning bolts. The investigating committee consisted of Maddox, James "The Amazing" Randi, and Walter Stewart. A theoretical physicist by training, Maddox had edited *Nature* for many years. Under Maddox, *Nature* has also been willing to express editorial views, often trenchantly. The second member of the investigating team, Randi, was an American professional magician who had become well known for his crusading efforts to expose and debunk pseudoscientists. He has a standing but as yet unawarded offer of $10,000 for the first person to demonstrate, to his satisfaction, the existence of some paranormal phenomenon. It is Randi's often-stated opinion that scientists (and especially physicists), far from being hard-nosed and skeptical observers, are extremely gullible. He demonstrated this theorem very clearly in the case of the Israeli spoon-bender Uri Geller after Geller had bamboozled a well-known theoretical physicist, among many others. Randi has himself exposed other instances of sloppy experimenting and has asserted that a professional magician should be a member of any investigating team where there is any possibility of deception, deliberate or otherwise, and especially when looking into parapsychological claims.

Stewart, the third member of the team, was a scientist at the National Institutes of Health (NIH) and became known outside his research field in 1972 when he refereed a paper for *Nature*. That paper reported on experiments in which rats were trained to avoid the dark. Material extracted from the brains of those rats was injected into other rats, which (it was claimed) then showed the dark-avoidance condition. That short report from scientists at Baylor School of Medicine and the University of Houston had less than a page of text. It was immediately followed by Stewart's seven and a half pages of criticism. (The claimed results have never been replicated.) Stewart has since turned into a scientific sleuth intent on rooting out and exposing scientific error and fraud. More recently, he was on leave from NIH to assist the staff of the congressional committee looking into allegations of fraud brought against scientists working with David Baltimore. (Baltimore, a Nobel Prize winner, was director of the Whitehead Institute at MIT. The congressional hearings are described in Chapter 10.) Stewart's forte is the statistical analysis of experimental data, but he did not have personal laboratory experience in the techniques employed by Benveniste and his group.

Nature's team of investigators spent a week in the Paris laboratory. Their four-page report appeared in the July 28, 1988, issue of *Nature,* together with an angry rebuttal from Benveniste. The investigative findings can be summarized by their section headings:

- The care with which the experiments reported have been carried out does not match the extraordinary character of the claims made in their interpretation.
- The phenomena described are not reproducible, but there has been no serious investigation of the reasons.
- The data lack errors of the magnitude that would be expected, and which are unavoidable.
- No serious attempt has been made to eliminate systematic errors, including observer bias.
- The climate of the laboratory is inimical to an objective evaluation of the exceptional data.

Apart from procedural aspects, the team was particularly critical of observer bias. This is a well-known problem that comes up when human observers have to decide what to count and even judge the precise value of a measurement, such as when comparing an object to a scale in an eyepiece of a telescope. Stewart's statistical analysis showed that the spread of some of the Paris laboratory measurements was far narrower than the distribution of values normally encountered.

"We conclude," the team said, "that there is no substantial basis for the claim that anti-IgE at high dilution ... retains its biological effectiveness, and the hypothesis that water can be imprinted with the memory of past solutes is as unnecessary as it is fanciful."[15]

The substance of Benveniste's response was predictable, but not his style. He defended the integrity of his co-workers and the validity of their results, and he directed his sharpest remarks at the investigating team and its methods, calling their work in his lab "a tornado of intense and constant suspicion, fear and psychological and intellectual pressure unfit for scientific work. ... I now believe this kind of inquiry must immediately be stopped throughout the world. Salem witch-hunts or McCarthy-like prosecutions will kill science."[16]

The investigators observed tests in which lab workers evaluated the degranulation in specimens whose identity and degree of dilution had been coded. To prevent any influence on the basophil counting, Randi sealed the master list of sample identification in aluminum foil, placed it inside an envelope, and taped it to the laboratory ceiling. Benveniste described Stewart's "typical know-it-all attitude" and said that he and Maddox had to "ask Stewart not to scream."[17] If we accept the accounts of the investigation, Benveniste's outrage is understandable.

Through four months, the correspondence columns of *Nature* carried many letters expressing all sorts of views, some critical of the journal's handling of the paper and the investigation. Some letters reported unsuccessful attempts to replicate the Paris findings. One of these came from Henry Metzger and Stephen Drakin at NIH.[18] Metzger had been a referee for the Benveniste paper, and he maintained that the paper should not have appeared.

The *New Scientist,* an independent British scientific weekly devoted to reviews and commentary but not to original research reports, allowed itself the unusual freedom of publishing an editorial it titled "Inhuman Nature." The piece expressed, with regret, opinions highly critical of *Nature* and its defense of its right to publish controversial papers: "Sadly, that venerable scientific organ has, in recent months, developed symptoms not unlike those often associated with rabies; madness and foaming at the mouth. ... Our complaint is in the way that *Nature* said it and the way in which the journal has responded to criticism of its behavior."[19]

In October, after providing Benveniste another page for his own commentary, *Nature* took four pages to respond and then ruled that its "correspondence columns should now revert to their normal uses."[20]

With all of the fuss that Benveniste's report generated, his explanation in terms of a molecular memory for water was risible. The water we use today has had a long and interesting history. Unless it has been synthesized in the laboratory from oxygen and hydrogen, its molecules have probably been recycled through oceans, clouds, rainfall, and rivers as well as through animals and people with who knows how many substances dissolved along the way. Why should not *all* of the dissolved substances have left their molecular footprints in the water that Benveniste used? Why should only the most recent memory be retained even after almost infinite dilution? Or perhaps Benveniste's water molecules, like Velikovsky's ancient societies, suffer from a collective amnesia of those early molecular encounters? The plausibility of both the results and the suggested mechanism can be gauged from the absence of any rush to duplicate those experiments.

The Benveniste affair has been short, very sharp, and entertaining in a warped sort of way. What are the issues, and what have we learned about the conduct of science? First, should *Nature* have published the original paper? It is a regular practice for journals to ask their authors to reconsider and resubmit a manuscript after making changes suggested by the referees and editor. Given the nature of the results being claimed in this manuscript, it was not unreasonable for *Nature* to ask for confirmation by independent laboratories. (The assisting laboratories turned out to be not fully independent, having been selected by Benveniste and having ties to homeopathy.) If the referees have not identified an obvious fault, unbelievability alone is a poor reason for rejecting a manuscript. The origin of the work and the reputation of the investigators are critical elements in the decision on publishing. Thus Maddox argued cogently in favor of publication:

> One of the purposes that will be served by publishing the article will be to provide an authentic account of this work for the benefit of those, especially in France, who have gathered rumors of it from the popular press. Another is that vigilant members of the scientific community with a flair for picking holes in other people's work may be able to suggest further tests of the validity of the conclusions.[21] And later: "It is rare that some such claim should come from a government-supported laboratory, that its principal author should urge publication in the face of common sense—and should complain that failure to publish will be tantamount to the suppression of the truth."[22]

In his final commentary Maddox set out the sequence of events and his reasons for various decisions made along the way, and his account is largely persuasive. He gave plausible reasons for the choice of members of his investigating team, pointing to their credentials, but he did not say why other investigators might not have been chosen. To his own question, "Has it been worthwhile?" Maddox replied, "The short answer is 'No.' Too many people have been caused too much distress, particularly at INSERM, and too much time has been consumed by the circus when there have been better things to do. But again the longer answer is that there are several ways in which the affair has been instructive, certainly for those involved and, it must be hoped, for many readers."[23]

Finally, we should raise the basic question, Has any other scientist bothered to attempt to replicate the experiment of the Paris group? The answer is yes: a group of scientists at University College, London, working with a statistician, followed "as closely as possible the methods of the original study" and reported that they could find "no evidence for any periodic or polynomial change of degranulation as a function of anti-IgE dilution." They went on to note that their "results contain a source of variation for which [they] cannot account, but no aspect of the data is consistent with the previously published claims."[24]

Conclusion

In Chapter 6 I described a number of radical proposals that were taken seriously by the appropriate scientists. Some of these ideas have now become permanent parts of science; others are still being tested; all have stimulated research. The contrast between those cases and the two examples I used in the present chapter begin to illustrate the transition away from mainstream science.

There are many similarities between the polywater and cold fusion cases. Both attracted a lot of attention by scientists who saw a need to test the reported results, to try to replicate the phenomena. Neither case has added to our fund of usable science (apart from presenting cautionary lessons). Both point to the difficulty in tracking down the causes of anomalies.

Benveniste's results stand isolated. They did not generate a stampede as did the cold fusion claim. Why should the collective reaction of the scientific community have been, in essence, a gigantic yawn? Right from the start, the claim for the powers of Benveniste's dilute solution was greeted by a skepticism qualitatively different from that attending polywater, relativity, or quantum ideas. Recall that polywater started as "anomalous water" and the initially reported anomalies seemed well-enough established to warrant further work. Relativity and quantum theory, although unpalatable to many, did not arrive without obvious connections to established science. Both resolved some puzzling anomalies, though it took many years for the sweep of their successes to be seen.

Why, then, did Benveniste's report fall with a dull thud? Probably because accepting its findings would have meant throwing overboard some cherished fundamental ideas, ideas that had been so well confirmed as to seem unassailable. Our modern science is based on linking identifiable causes with demonstrated effects. When the *absence* of a causative agent is linked to a claimed effect, it should be no surprise that the theory meets with general disbelief. We see no way in which the memory of the prior presence of dissolved molecules can be retained.

The reception to Benveniste's paper was in marked contrast to the nonsensation after a paper reporting a somewhat similar effect had appeared in *Lancet* in 1986. "Is homeopathy a placebo effect?" the authors (from the Glasgow Homeopathic Hospital) had asked, and after carrying out tests, they concluded that "no evidence emerged to support the idea that the placebo action fully explains the clinical responses to homeopathic drugs."[25] It was the appearance of Benveniste's paper in *Nature*, with its wider circulation, and with the editorial comments, that drew the attention.

Scientists' deep belief in causality was being challenged by the results from Paris. It is probably fair to say that if (a very big *if*) Benveniste had been correct, we would have had to wonder about the validity of the entire structure of science. Alternatively, nature would have to be seen as capricious, sometimes linking cause and effect, at other times not. Conventional science has been so successful that we are unlikely to abandon its major premise, but can we exclude an occasional, random blip in which the "impossible" does happen? If so, this would demand a new definition of science, something that is very far from being forced on us on the basis of the present evidence. And so Benveniste's results were ignored by just about everyone other than *Nature*, and about five years elapsed before there was any attempt to replicate. In 1993 the National Institute for Medical Research decided to close Benveniste's laboratory following a regular procedure in which research support is deliberately moved around.[26] What will our reaction be if Benveniste or someone else repeats his measurements and obtains positive results? My guess is that we will probably see a few other experiments, tentative at first, until there is a much clearer pattern of confirmation. However, unlike the polywater case, the original results raised such fundamental skepticism that further work seems unlikely, though the popularity of the homeopathy industry may ensure continuing interest.

Notes

1. Brian A. Pethica, *Journal of Colloid and Interface Science* 88 (1982): 607.

2. *Polywater* (Cambridge: MIT Press, 1981). Pethica (n. 1) wrote a hostile review of this book.

3. F. J. Donahoe, in *Nature* 224 (1969): 198.

4. *Scientific American* 223 (November 1970): 52.

5. *Polywater*, 119.

6. *Nature* 333 (1988): 816. IgE is Immunoglobulin E.

7. Ibid., 818.

8. *New Scientist* 119 (July 14, 1988): 39.

9. Ibid., 119 (August 4, 1988): 30.

10. *Nature* 334 (1988): 372.

11. *New Scientist* 119 (July 14, 1988): 39; 119 (August 4, 1988): 31.

12. *Lancet* (October 18, 1986): 881.

13. *New Scientist* 119 (August 4, 1988): 31.

14. *Nature* 333 (1988): 787.

15. Ibid., 334 (1988): 287.

16. Ibid., 291.

17. Ibid.

18. Ibid., 375.

19. *New Scientist* 119 (August 18, 1988): 19.

20. *Nature* 335 (1988): 759, 760.

21. Ibid., 333 (1988): 787.

22. Ibid., 335 (1988): 760.

23. Ibid.

24. Ibid., 366 (1993): 525.

25. *Lancet* (October 18, 1986): 881.

26. *New Scientist* 140 (1993): 10.

8

Tabloid Science

Fads and Fallacies in the Name of Science is Martin Gardner's classic 1952 omnibus of many of the reports, theories, and "insights" that sadly deluded and confused people have considered to be genuine and scientific. Gardner's book is still in print and in spite of its age provides a wonderful introduction to this genus of pseudoscientific pathology. It is both entertaining and depressing to read through Gardner's case histories—entertaining for his thorough documentation of the errors of the would-be savants and depressing because more than forty years later so little has changed. There are new generations of authors bravely rediscovering the same old fallacies in the pseudoscientific landfill; they use the same old methods to present and semidocument their theories, the same old pseudoscholarship that fills the pages of too many books and magazines and the hours of endless late-night talk shows.

The variety of such examples of pseudoscience is so wide that in a single chapter I can do no more than present a selection to demonstrate some topics and methods. The first topic, astrology, is by far the oldest, though never previously as profitable as it now is. The phenomenon of unidentified flying objects is of more recent origin and seems likely to persist at a low level. Much of this sort of fringe science tends to be episodic, its memory soon fading and its brashly promoted books quickly going out of print. (Fortunately or unfortunately, I have continually had difficulty in trying to assign these books to my students, for, as it is so tersely phrased in England, today's news wraps tomorrow's fish.) There is an inexhaustible supply of new nonsense to replace the old. As you will see, some of the examples I have chosen have vanished, but others are the soap operas of pseudoscience—thin of plot and predictable in outcome, limited in character development, but still money-makers.

Astrology

There is no way of escaping the obligation to deal with this subject so often debunked yet still with us. However, when astrological signs gain credibility in the Oval Office of the White House, we have a problem of a different dimension from any we have yet encountered. Scientists have, in our superior way, scorned astrology, passing only occasionally to make a condescending comment regarding those poor confused people who put their trust in its predictions. We acknowledge that hundreds of years ago and before we had our present understanding of the laws of science such as those for gravity and electromagnetism, there was no reason not to suspect that events on earth were at the mercy of the location of the sun, moon, and other heavenly bodies. Astrologers were appointed by many courts, their advice sought before the emperor signed a treaty or gave the hand of his daughter in marriage or entered battle. For 2,000 years, until 1911, there was an official court astronomer in China whose obligation was the careful recording of all sorts of heavenly apparitions, omens that might have good or bad meaning. Transient objects were especially noted and called "guest stars." These records have proved to be of great value to modern astrophysicists; the guest stars include comets, meteors, and exploding stars, and the observational records are very reliable. Almost every return of Halley's comet has been noted since 239 B.C.

But then, as we thought, scientific discoveries had led to enlightenment, an understanding of the roles of the sun and moon in affecting the tides and weather on earth. The constellations were seen for what they really were: chance groupings of stars, often at vastly different distances, in patterns with arbitrarily assigned boundaries, convenient as signposts in the skies, sufficiently remote to be admired, but without earthly influence. There was no longer any reason to suspect that the appearance of a planet against any particular backdrop of stars could have significance, could possibly effect or foretell events on earth.

But commercial astrology is now a big business. Syndicated columns of predictions appear in many (too many) newspapers. Some of these carry a disclaimer, such as "Horoscopes have no basis in science. They should be read for amusement only." This is as honest and effective as the warnings on cigarette packs. A syndicated astrologer who receives only a few dollars from each newspaper but whose horoscopes appear in 100 papers each day soon has a fine income. There are over 10,000 astrologers in the United States; check your yellow pages for the one nearest you.

The essence of today's astrology is that knowledge of the positions of the sun, moon, and planets at special times can be used to construct a horoscope for each person, spelling out that individual's personality and indicating when conditions are propitious for major decisions or when the heavens foretell some important event. The "special times" can refer to the time of birth (sometimes the time of conception, which clearly introduces some interesting uncertainties); the "positions" can be rising or setting, or at noon or midnight, and against which constellation (or astrological "house") the heavenly body is located at that time. The constellations used in astrology were named in Babylonian times, but the astrological signs that bear the same names today are no longer coincident with them. The twelve houses are sectors of the sky, located relative to the horizon. The first house is the sector just below the eastern horizon; a planet in this house will rise within the next two hours. The second house contains objects that will rise two hours later, and so on.

"Predictions" are usually phrased in such vague terms that there is great elasticity in interpreting a horoscope, something for everyone to identify with. Commercial astrologers seem to have been generally unwilling to undertake research and produce systematic measures of their effectiveness—or have not done so to the satisfaction of scientists. Skeptics, curious bystanders, and a number of astronomers have, however, investigated the astrological claims. Without exception, when critically examined, astrology has struck out, its predictions at best no more accurate than random guesses. Where successes have been claimed, there is sufficient flexibility in the forecasts that firm believers can always find a way of wriggling out of the corner. Michel Gauquelin, a French psychologist who became intrigued by astrology, carried out some simple tests with fascinating results. Responding to an advertisement by a commercial astrologer, he submitted birth data for ten notorious criminals without identifying them as such. The resulting horoscopes were benign, giving no glimpse of the true character of those murderers. He then himself advertised that he would supply free horoscopes. To the many people who sent in requests, he provided copies of the horoscope of a murderer. Most of Gauquelin's respondents claimed to see their own personalities described in the horoscope. There have been other tests of this sort, and the results show how prone we are to self-deception. Again, Gauquelin undertook perhaps the most extensive statistical examination of horoscopes. Puzzled by the survival of astrology over so many hundreds of years if it truly had no content, no predictive power, he analyzed the horoscopes of many thousands of people. He could find no relationship between their lives and the positions of the planets in the various astrological signs or houses at the times of their birth. He did, though, stumble onto a strange effect that I describe later.

A particularly striking test of natal charts was carried out in the mid-1980s at the University of California in Berkeley by Shawn Carlson, with the assistance of statisticians, other scientists, and astrologers. They drew up experimental protocols to incorporate "all suggestions by the astrologers provided they could be followed without biasing the experiment."[1] In the tests astrologers were asked to identify subjects from three sets of profiles, some based on natal charts and some from the widely used California Personality Inventory (CPI). Subjects were also asked to identify descriptions of themselves from the same set of profiles. The conclusions were twofold. The astrologers performed no better than expected on the basis of chance. This may be no surprise to skeptics, but the second finding was even more interesting. It emerged that the subjects could not even "recognize accurate descriptions of themselves at a significant level."[2] What this implies is that one cannot accept at face value the repeated claim from so many people that astrological charts provide remarkably accurate descriptions of themselves. This controlled test confirms the findings of Gauquelin.

Surveys continue to show a public ambivalence toward astrology—a measure of hesitant disbelief coupled with a feeling that perhaps there really might be something in it after all. All of this fuels a cottage industry of comfortable proportions, for in addition to the daily newspaper columns, there are personal consultations, seminars, books, and so on. The late George Abell, a UCLA astronomer, estimated that the expenditure on astrology runs into billions of dollars each year. Generally, we might consider this wasteful but harmless, certainly less cruel a pseudoscience than the claimed cancer cures. Our sense of superior amusement should have been brought up short in May 1988 when we learned that President Ronald Reagan's schedule was sometimes adjusted because of astrological forecasts. We do not yet know the full extent of this celestial influence, and we can discount any claims from Joan Quigley, former under secretary for astrology; it is bad enough that *anything* hinged on her prognostications. Confirmation of the intrusion of astrology came from Donald Regan, chief of staff to Reagan, who has written that "virtually every major move ... was cleared in advance."[3]

Perhaps Reagan had not read the statement that was drawn up in 1975 in an attempt to discredit astrology. "Objections to Astrology" was first published in *The Humanist* and then widely distributed.[4] The intended purpose of the statement was to indicate that many scientists and scholars found no basis for belief in astrology. The authors wanted to counter the usually open-armed reception that astrologers receive throughout the media, as they are welcomed onto talk shows and given space in print usually without any balance in presentation. The scientists hoped that through this statement the general public might at least become aware that astrological predictions were not as reliable as claimed.

Prime movers behind the statement were Bart Bok, an astronomer of international standing, and Paul Kurtz, professor of philosophy at the State University of New York at Buffalo. They circulated the statement, mostly among members of the National Academy of Sciences, and published it with 186 signatures, many from Nobel Prize winners.[5] Out of this concern with the uncritical public acceptance of astrology came the founding of the Committee for the Scientific Investigation of Claims of the Paranormal (CSICOP), with its *Skeptical Inquirer,* a most useful quarterly publication.

The statement opened, "Scientists in a variety of fields have become concerned about the increased acceptance of astrology in many parts of the world." It went on, "We, the undersigned ... wish to caution the public."[6] This probably did no good; it sounded too much like the shopworn "nine out of ten doctors recommend." The risk of issuing such a statement was clearly shown when Paul Feyerabend, a maverick antiscience philosopher, responded.[7] He compared the statement to the *Malleus Maleficarum* issued by the Catholic church in 1484 as a warning against witchcraft. Pope Innocent VIII provided an introductory bull, declaring that "it has indeed come to our ears, not without affecting us with bitter sorrow, that ... many persons ... have strayed from the Catholic Faith and abandoned themselves to devils." Feyerabend, noting the close parallels between the 1484 and 1975 statements, was not defending astrology but, as he often does, was trying to deflate the scientists. One lesson that can perhaps be drawn from his essay is that scientists need to take greater care when making public pronouncements. Feyerabend notwithstanding, conventional astrology has, to scientists, been discredited, but there is a related phenomenon that deserves a brief mention.

We have already encountered Gauquelin, who carried out an investigation into astrological predictions. Although he could find no evidence to support the claims of conventional astrology, he did turn up something totally unexpected when he looked into the birth times of famous or very successful people: a correlation between their times of birth and certain planetary positions in their daily apparent orbit around the earth, unrelated to the constellations. Because of the earth's rotation, each planet seems to make an orbit around the earth once each day. Gauquelin divided each apparent orbit into twelve sectors, with the first denoting the two-hour interval immediately following the rise of that planet. If planetary positions had no influence at all on our lives, then one would expect just as many births to occur for planetary positions in each of the twelve sectors. Instead, for a number of groups of very successful people, Gauquelin found that a disproportionate number had been born at times when Mars or Jupiter or Saturn had been in the first or fourth sectors. One of his earliest samples of 576 doctors was based on members of the Académie de Médecine, and they had "chosen to come into the world much more often during roughly the two hours following the rise and culmination of two planets, Mars and Saturn."[8]

Gauquelin was joined by his wife, Françoise (also a psychologist), and together they extended his studies, examining other groups of prominent people: For "sports champions" and "exceptional military leaders," Mars turned out to be the indicator; for actors and politicians, it was Jupiter; and for scientists, it was Saturn (which was avoided by painters and musicians). The basic form of analysis seems completely straightforward. Using birth records, one can count up how many famous people were born when any particular planet was in each of the twelve sectors. Standard statistical tests can then be applied.

As one would have expected, the apparent simplicity of the procedure invited checking. The Comité Para, composed of scientists in Belgium, conducted the first tests. They initially agreed with Gauquelin's calculations for athletes, then found that the correlation with the positions of Mars vanished if athletes born in Paris were excluded. This sort of volatility seems to be a persistent pattern in that the results appear to be sensitive to the method of selection of the data base. In the United States tests were carried out by George Abell and Paul Kurtz, both affiliated with the CSICOP, and Marvin Zelen, a Harvard statistician. They used the Gauquelins' data base as well as those that they themselves had generated, and they concluded that the data were "consistent with an equal chance of birth times falling within each of the twelve sectors" and that "the analysis of American sports champions shows no evidence for the Mars effect."[9]

There has been disagreement over the selection of people to whom the test can be applied. Directories such as *Who's Who* have been employed, along with membership lists for prestigious academies and names of athletes who have appeared in all-pro or all-star teams. But not everyone on each list gets selected for the analysis. For example, in one test the Gauquelins rejected basketball players, "for they have given the most disappointing results." The Gauquelins also found that the Mars effect did not show up for soccer players in the Italian first division who had not played on international teams. The Gauquelins also criticized the way in which the Kurtz group made their selections.

There was an exchange of analyses in the pages of the *Humanist* and in the *Skeptical Inquirer,* each side maintaining that it had found or disproven the claims of the other and each criticizing the other for the statistical analysis used and the decisions to include or exclude subsets of the data.[10] As a scientist, I find the selection criteria of the Gauquelins far too subjective to be trustworthy. In his books Gauquelin has provided colorful detail of the exchanges he had with his critics. Michel Gauquelin died in 1991, but the debate has continued. Suilbert Ertel, a psychology professor at the University of Göttingen, undertook a sophisticated statistical study of Gauquelin's data and tried to make estimates of the "eminence" in a quantitative way.[11] He found some positive correlations but, surprisingly, also found some anticorrelations, that is, a planetary avoidance of some zones at the birth times.

There is a total absence of any known scientific effect that could link planetary positions (at *any* time) with fame or career success, but this is not a conclusive reason for the rejection of the Mars-effect hypothesis. Discoveries often precede theories. But in case you think that this is an equal standoff and that the Mars effect has some scientific credibility, I must comment that this is not the way that science advances. The burden of proof rests primarily on the person claiming the discovery. In this case enough doubt has been cast on the reality of the Mars effect that most scientists will probably not bother with it, unless a far more significant effect is convincingly demonstrated. Certainly, the Mars effect is not widely known, and there has never been a long line of scientists waiting to undertake their own checks. At the same time the exchanges between Gauquelin and his critics, between commercial astrologers and their debunkers, seem to have made no impact on the popular acceptance of astrology. Groups of enthusiastic debunkers and local groups of skeptics beaver away, but all to no lasting effect. Why bother? I return to this central question in Chapters 13 and 14.

I cannot leave the subject of astrology without taking note of a strenuous defense of the scholarly study of astrology, an assertion of the importance of the role of astrology in the history of science. Otto Neugebauer, for many years the dean of historians of Babylonian astronomy, published a short note in *Isis,* the publication of the History of Science Society (HSS). George Sarton, HSS founder and the first editor of *Isis*, had in an earlier review referred to a book as "a wretched collection of omens, debased astrology and miscellaneous nonsense."[12] Neugebauer pointed out that astrology had played an important role in the development of the sciences—"No Arabic astronomer can be fully understood without a thorough knowledge of astrological concepts"—and he objected strongly that "the recognized dean of the History of Science [would dispose] of a whole field of knowledge with the words 'the superstitious flotsam of the Near East.'"[13] Neugebauer was correct. Historians could do useful work on astrology's role in intellectual history; today, however, the study of astrology's claimed effectiveness is surely a waste of time for astronomers (though a more fertile field for psychologists).

The Jupiter Effect

Mainstream astrology has for centuries been obsessed with the positions of the planets, moon, and sun and their implied influences on human affairs. John Gribbin and Stephen Plagemann used a different approach for their 1972 book *The Jupiter Effect* by adding a whiff of real science based on space-age discoveries. There was some solid scholarship in this book but in places so far behind that the connections got tangled. What truly distinguishes this book is that Gribbin, with his Ph.D. in astrophysics from Cambridge University, is a fine popularizer and a regular contributor to the *New Scientist*.

The essential point of *The Jupiter Effect* is that on rare occasions (at intervals of 179 years) the sun, earth, and planets line up, producing a maximum gravitational force on the sun as well as on the earth. This force raises a tide on the sun, much as tides are raised on the earth by the gravitational pull of sun and moon. Taking this minuscule effect on the sun, Gribbin and Plagemann went on to assert that the solar tides affect sunspots. The sunspots, in turn, release high-speed particles that stream out through the solar system to affect the earth by changing its rotation speed and so producing earthquakes. The language these authors used in setting out their prediction carried the weight of scientific authority and was unambiguously emphatic: "Now we can predict this apocalyptic date to within a couple of years. ... In 1982 ... Los Angeles will be destroyed."[14]

Gribbin and Plagemann arrived at this book-selling conclusion through consideration of an interesting collection of solar and planetary observations and events. They assembled their case methodically, unveiling their prediction for 1982 as the inevitable result of uncontrollable cosmic processes. Their starting point was a review of modern ideas of plate tectonics and earthquakes, then a description of the San Andreas fault in California and the 1906 San Francisco earthquake. They introduced the astronomical side with material on changes in the earth's rate of rotation and its possible association with the weather. Then they described the solar wind because of its possible influence on the earth's rotation. (The solar wind is a stream of electrically charged particles continually flowing out from the sun, past the earth, and out through the solar system.) Other potentially influential factors they considered included solar flares and the intense bursts of particles they eject with consequent effects on the cosmic rays coming near the earth. Because my research interest has long been in cosmic rays, I am familiar with many of Gribbin and Plagemann's chosen topics. I found their writing interesting, but a careful reading and reference to the cited literature show that the logical development needs watching—the style of the argument is too often innuendo: "*If* the Sun's activity ... *then surely*" (emphasis added).[15]

The eleven-year cycle in the number of sunspots is well known, and Gribbin and Plagemann covered this territory that is so familiar to cosmic-ray workers. They discussed at length a theory that relates sunspot activity to the tidal height on the sun.[16] In a table of data taken from a paper in *Nature* by K. D. Wood, they listed the dates of the peaks in the sunspot numbers and the dates of the tidal peaks (as calculated from planetary positions) from 1604 onward.[17] The correlation is impressive. It is most unfortunate that in 1972 Gribbin and Plagemann could not take into account a 1976 paper by John Eddy, who showed, among other things, that almost *no* sunspots were seen between 1650 and 1715.[18] (This lean period is now known as the Maunder Minimum, and its physical basis is a source of considerable puzzlement to astrophysicists.) Galileo's discovery of sunspots came in 1610, but sunspot data were sparse for several years thereafter, so that one must also wonder about the reliability of Wood's designation (and Gribbin and Plagemann's acceptance) of 1604 for a sunspot maximum and for other data listed during the Maunder Minimum.

The next links in their chain were the influence of Mercury on the number of sunspots, then the planetary alignments (easily predicted these days), and the expected 1982 alignment, "a superconjunction with all the nine planets in line on the same side of the Sun."[19]

As with Velikovsky's books, Gribbin and Plagemann's work includes much documentation; references are scattered through the book and take up the final eight pages, with many citations from the professional literature: the *Bulletin of the Seismological Society of America*, the *Journal of Geophysical Research*, the *Monthly Notices of the Royal Astronomical Society*, as well as books and articles by well-known scientists.

Of course Los Angeles was not destroyed in 1982, nor had it been by the time of this writing. Still, San Francisco was severely damaged in October 1989, just at the scheduled starting time of the first game of the World Series. Unfortunately, that was in a year Wood (and Gribbin and Plagemann) had projected to be well between two solar tidal maxima. Reference to the *Astronomical Almanac*, published each year by the U.S. Naval Observatory and the Royal Greenwich Observatory, shows that the planets did not even line up at the time of this major earthquake. In fact, the directions from the sun to Venus and Jupiter were not aligned but formed a right angle. Wood, in his *Nature* article, had noted that only Mercury, Venus, earth, and Jupiter made significant contributions to the solar tide and that none of the others raised more than 2 percent of the earth's contribution. The requirement that *all* nine planets line up was therefore dramatic but unnecessary. What had gone wrong? Lots, as you will have guessed.

In his review of the book in the *American Scientist*, published by the respected scientific honorary society Sigma Xi, Don L. Anderson, a Cal Tech seismologist, was caustic: "The book ... is a commercial commodity of the crassest kind. The authors are clearly after the large cult and astrology market and the many Californians genuinely fearful of earthquakes."[20] Pointing out that "seismologists and meteorologists are currently working on the problem of the ultimate cause of irregular changes in rotation" of the earth, Anderson went on to say that "an impressive amount of evidence has been assembled, but none of it is to be found in this book."[21] Perhaps the most devastating criticism, because of its great detail, came from the Belgian astronomer Jan Meeus in a lengthy paper in *Icarus*, a professional journal devoted to solar system studies. Meeus refuted Gribbin and Plagemann point by point, leaving their assumptions and theory shredded. Their documentation of sunspot numbers on the east and west limbs of the sun was wrong. There is no dependence of sunspot numbers on the position of Venus: "Any speculation about planetary tidal effects on the Sun and on its activity is completely irrelevant."[22] The planets would not be lined up in 1982. And so it goes on, with Meeus demolishing point after point of *The Jupiter Effect*.[23]

But the hypothesis of the potency of tidal forces remains attractive, and it emerged again in 1990, minus the sunspots, when a major earthquake was predicted for Missouri. This fantasy is the subject of the next section.

"Earthquake—or Earthquack?"

This was the heading for a news report in *Science*.[24] Between June and December 1990, a wonderfully pseudoscientific episode unfolded. It has been very well documented, and its repercussions were extensive. A prediction was made and widely publicized that there was a fifty-fifty probability that a major earthquake would occur along the New Madrid fault within a day or two of December 3, 1990. The region at risk was centered on southeast Missouri but extended north to include the area around St. Louis. The forecast came from someone with no credentials in seismology, but, despite authoritative critiques, it stimulated a near panic among some people. December 3 came and went without even a coffee cup's being spilled from geophysical causes. Normally, a nonevent is scarcely noteworthy, but this one was different. (As a friend has put it, who would read a story with the headline, "Man Shaves Himself Safely, Once Again Escapes Slitting His Throat?")

The earthquake prediction came from Iben Browning, whose most recent entry in *American Men and Women of Science,* in 1971, listed his Ph.D. as being in zoology.[25] He cited his professional interests as including zoological topics, aeronautical electronic instrumentation, genetics, optical engineering, information theory, and neural nets. For a number of years, Browning was an independent consultant for the investment firm of Paine-Webber, among others. Browning died in July 1991.

In October 1989 Browning's newsletter gave the first of many warnings of impending geological events, predicting volcanic activity in the Northern Hemisphere.[26] In a speech for Paine-Webber, he narrowed the geographical zone to 30°–45° north latitude. In a later talk in Kansas City in December 1989, Browning specified the New Madrid region, and he repeated this assertion in his December newsletter. Later a videotaped interview was marketed in two versions, one selling for $39 and the other for $99. "Should you take Dr. Browning's projection seriously? Should you and your family take steps now to prepare?" asked an advertisement.

Was there any scientific basis for Browning's prediction? As with the better pseudoscience, there was a grain of plausibility. Earthquakes are associated with movement of giant tectonic plates in Earth's crust, and plate boundaries are the sites of 95 percent of all earthquakes. The New Madrid fault, well known to seismologists, is unusual in having a very deeply buried fault, described as a "failed arm of a crustal rift."[27] This is the second most seismically active region of the continental United States, though most of its events are small and pass with little or no notice. However, there was a period of major activity between December 1811 and February 1812, when New Madrid was the site of three strong earthquakes. From descriptions of the effects, it has been estimated that these three events had magnitudes exceeding seven on the Richter scale. Since then, there have been only two events along that fault above magnitude six, in 1843 and 1895. In September 1990 there was an earthquake of magnitude 4.6, which of course involved several thousand times less energy than a magnitude seven event.

There is professional consensus that a major earthquake (of magnitude six) will probably occur in the New Madrid region during the next thirty years, but it is not expected to produce significant damage in St. Louis. An earthquake of magnitude seven is far less likely. Despite extensive research, the prediction of earthquakes has not yet attained the status of a reliable science. Many plausible hypotheses and possible warning signals have been examined, but none has yet been shown to be a reliable predictor. Browning built on one of these discarded ideas. The gravitational pull on the earth caused by the sun and moon varies in a complex cyclic way as the relative positions of these bodies change. There are smaller effects arising from the noncircular shapes of the orbits and also from the presence of the other planets. The part of the earth's surface that feels these variable tugs is continually moving as the earth itself rotates. The best-known result of this gravitational attraction is the movement of the oceans, the tides. The gravitational forces that make the waters move also produce tides in the solid part of the earth, but only on a minuscule scale. Browning computed the total tidal force on the earth and how it changes over the years. His theory was that earthquakes are most probable at times of greatest tidal stress, when the release of a large amount of stored energy can be triggered.

When Browning's prediction began to gain wide publicity, in summer 1990, seismologists at St. Louis University and Washington University issued statements pointing out that this seismic model had been examined extensively but that no demonstrated correlation could be found between tidal force and earthquakes. The U.S. Geological Survey (USGS) assembled an ad hoc working group that studied Browning's forecast and issued a report that shredded his method, prediction, and claims for prior successes.

In support of his claim that tidal force analysis does show a correlation with earthquakes, Browning had quoted a paper that appeared in 1968 in *Icarus,* a respected professional journal.[28] Unfortunately for Browning, in 1970 that finding of a positive correlation was shown to be based on a mathematical error, but Browning did not refer to the later paper.[29] The ad hoc group dryly noted that

perhaps the most common problem with published studies of the correlation between earthquakes and tides is a misuse of statistical methods. Having well defined rules is essential if statistics are to have meaning. The statistics are invalid if results are examined before deciding whether or not to play the game. Although this mistake is commonly made in tidal triggering studies, it is clear that professional casinos do not allow their patrons to make this same mistake.[30]

Just as unfortunate for Browning's method, major earthquakes have occurred when the tidal forces were at their lowest values or between maxima, and they have failed to appear at times of maximum force. Browning's calculation of tidal force does not lead to a specified location on the earth's surface where the effects will be felt. For December 1990, he started by predicting events within a broad band of latitudes such as between 30° and 45° or 60° north. This is not a usefully narrow prediction since on average there are just over 100 earthquakes above magnitude six in this latitude band each year. By some unrevealed method, Browning decided that the New Madrid region was at particular risk.

A friend who is interested in the pseudosciences has drawn my attention to a speculation regarding Browning's method of predicting. Consider the sequence of numerals 1 2 3 4 5 6 7 8 9 0. Divide this into groups: 12–3; 4:56; 7.8; 90. The month and day of the predicted earthquake is the 12–3, the time of day is 4:56, the magnitude of the quake is 7.8, and the year is (19)90.

In any event Browning claimed to have successfully forecast the 1971 earthquake in San Fernando, the 1972 earthquake in Nicaragua, and the California earthquake that so spectacularly delayed the start of the World Series in October 1989. Again unfortunately for Browning, the ad hoc group examined videotapes, a transcript of a talk he had given, and his newsletters. Browning stated that "there will probably be several earthquakes around the world" but made no mention of California.[31] The group's conclusion was that "Browning's correlations of earthquake activity with danger periods does no better ... than random guessing. ... Since there is no public record of publicly stated specific earthquake predictions that he made prior to 1989, we can only infer that successes claimed before that time were retrospective."[32]

If this was all that had happened, we could have chalked it up as yet one more dreary example of a discredited pseudoscientific prediction to go along with the Jupiter effect. After all, who bothers to keep track of the "predictions" of various commercial astrologers? But there is a difference. The daily astrological silliness is individually targeted; there are no apocalyptic communal predictions. Earthquakes, in contrast, are very real, and in 1990 the previous year's California event was still a vivid memory, after nationally televised coverage of collapsed buildings and bridges. The public was sufficiently sensitized that when a small quake occurred in September 1990, many people thought that this showed the correctness of Browning's technique. In fact, that quake took place at a time of minimum tidal stress, a matter of little consequence to a nervous public.

As December 3 approached, few people in the New Madrid area of the Midwest were untouched. Many local emergency management organizations took advantage of the inescapable drumbeat of publicity to alert residents to reasonable precautions they could take. In a seismically active zone, taking these precautions would have made some sense in normal times, but they could not escape being tainted by the growing hysteria. Schools held evacuation drills or practiced taking shelter under tables and desks. Individuals and groups laid in stocks of durable supplies. At the end of October, the St. Louis *Post-Dispatch* produced a well-designed, twenty-page "Earthquake Preparedness Guide." In an editorial William Woo, the editor, set out the reasoning behind publication of the guide. After noting that the experts disagreed with Browning, he commented, "What should the *Post-Dispatch* do in response to the developing public anxiety? To fuel apprehension and possibly set off a panic would be irresponsible. But what if we carried on as usual, making no special effort, and an earthquake occurred? The knowledge that we could have helped our readers—but did nothing—would be intolerable."[33] This is not an argument to be brushed aside, given that there was responsible seismological opinion for a major event sometime within the next thirty years and also considering the extent to which public concern had been aroused. The *Post-Dispatch* guide provided information on earthquake mechanisms and useful checklists of things to look for, of precautions to minimize damage and injury. All sorts of devices were advertised—gas shutoff valves, plastic film for windows, portable stoves, electric generators. The effect was somewhat diluted by the advertisement of one enterprising company; under the heading "earthquake survival kits," it listed "first aid kits, flashlights, handguns, military rifles, survival knives." (We had encountered similar ingenuity in the early 1960s, when fallout shelters were being pushed and the need to protect yourself against your unprepared neighbors was being advocated.)

New Madrid, a small town in southern Missouri, was invaded by national and foreign reporters unwilling to pass up the chance of being witness to the great event. There were live reports (I hesitate to call them action reports) on the national networks. Many schools announced that they would be closed over the projected danger period of December 3 and 4. Some parents sent their children out of the area. Even in St. Louis there were people whose veneer of sophistication suddenly evaporated as they made plans to sell their houses and move to more secure structures. Earthquake insurance was suddenly a topic of conversation and active commerce.

I am here to tell you that we all survived. For many of us who had been making patronizing statements to our students, our greatest fear was that there *would* be an earthquake on the chosen date. Even though we had been careful to point out that there was a real chance something would happen, though not necessarily on December 3, it would have been hard to maintain our credibility if there had been an earthquake on the announced date. Browning would have seemed to be correct in the face of the derision of the experts.

Through it all, the media behaved with impeccable impartiality, giving space and time to Browning and to his critics as though they were equally respectable. To a degree, the credibility of Browning's forecast had been enhanced by support from David Stewart, director of the Center for Earthquake Research at Southeast Missouri State University, but tough investigative reporting was notably absent. Browning's record and Stewart's expertise could and should have been reviewed by the media and not simply accepted. For example, in 1976, when there had been another questionable earthquake prediction (for North Carolina), Stewart had been aided by a psychic. Browning's unsubstantiated claims could have been easily uncovered, as the ad hoc working group had already done in October 1990, but by then the publicity had been extensive and the public was nervous. The naivete of the media in the handling of this episode has been revealing. One of the most interesting comments came on December 2 at a meeting organized by the St. Louis chapter of the Society of Professional Journalists. One panelist, a reporter for the St. Louis *Post-Dispatch,* noted that "Browning was a consultant to businesses, a research institution and other organizations. Still others subscribe to a newsletter published by Browning. I'm not prepared to say that all those people are stupid."[34]

And so memories of Browning and his unquake are fading. But should they? When Margot O'Toole alleged misrepresentation of her scientific data (see Chapter 10), there were several inquiries and a congressional hearing (the pretext for congressional involvement being use of public funds). We are unlikely to see a congressional hearing on the Browning case, though perhaps we should have had an investigation into why such a patently pseudoscientific and flimsy forecast should have been taken so seriously, its importance magnified, causing significant public inconvenience and expense. This is a real case of crying wolf. When a responsible scientific prediction is made at some time in the future, whether of an earthquake or some other impending natural disaster, will it be heeded, or will it evoke the response, "Remember Browning"? Premature warnings have long been a matter of concern to responsible scientists. For example, a study by the National Academy of Sciences in 1975 had warned against precisely this sort of scenario, the premature issuing of earthquake warnings based on unproven methods.[35] But in the public mind, aided by too many of the media, Browning's credibility was on a par with those of the professional scientists, disclaimers notwithstanding. It is true that the St. Louis *Post-Dispatch* did give extensive coverage to the report from the USGS ad hoc group, highlighting its critical comments. This comes perilously close to the sort of "balanced treatment" that creation scientists want for their particular cause. It is interesting to see that alongside the *Post-Dispatch* report and of almost the same length, there was a report on a conference of UFO believers.

The USGS has produced a useful compilation documenting much of the Browning episode.[36] In addition to a short analysis and commentary, this publication includes photo reproductions of Browning's newsletter, the texts of some of his talks, and about 170 pages of articles and editorials that appeared in various newspapers.

The need to make a distinction between science and pseudoscience is well exemplified by the Browning case. When the public cannot tell the difference and the media blur the distinction, we have a situation that can be serious. When we apply the various tests of Langmuir, Bunge, and Gardner (see Chapter 12) to the Browning case, we find some characteristic signs. Authoritative scientists were uniformly ranged against his method and prediction. Browning did not publish in a refereed journal. He worked alone, well outside the mainstream. He was resistant to criticism. When the USGS team invited his comments, he noted that "it would seem gratuitous to me for any seismologist to deny the enhanced probability of great earthquakes around rarely spaced high vector sum tidal forces."[37]

UFOs and Their Occupants

"I saw it with my own eyes" is not, regrettably, any guarantee of either reality or reliability. We often see what we expect or wish to see, and our imagination can intervene between the signals reaching our eyes and our later recollections. Psychologists have repeatedly made these points, as have lawyers confronting witnesses in countless court cases.[38] All of this is especially noticeable when we come to examine the topic of unidentified flying objects, or flying saucers, as they are often called.

Though the U.S. military had earlier received reports of strange flying objects, there seems to be a general agreement that the first public report came from Kenneth Arnold in 1947.[39] Arnold was flying near Mount Rainier when he saw a group of nine circular objects flying very rapidly, in formation, between the mountain peaks. In describing what he saw, he compared their movements to those of a saucer skipping across water. The rest, as they say, is history. Since that time there have been thousands of reported sightings. In some cases the "vehicles" had occupants. In extreme cases they have landed and even abducted the human witnesses, later returning them to recount their unbelievable adventures. Books, magazines, and newspaper stories continue to record yet more of these appearances.

The reports fall into several reasonably distinct categories. There are those like Arnold's, seen either from airplanes or from the ground, in which one or more objects move at high speed, usually in silence and performing maneuvers that are not possible with known aircraft. Sometimes flames or exhaust is reported; sometimes the apparitions are close enough for observers to note details in their shapes, some described as cigarlike, others as disks. In some cases radar echoes have been picked up showing something moving at high speed. In other cases glowing objects have been seen. In many of the reports, strange behavior has been noted in nearby animals. The accumulated evidence is far too persistent to be ignored, written off as simple fantasies. What is the reality underlying these reports?

Many can be explained easily in terms of well-known objects of natural though sometimes unusual phenomena. Inexperienced observers have often mistaken the planets Venus and Jupiter for spacecraft. After the sun and moon, these two planets are the brightest objects in the sky, and at its brightest Venus can even cast shadows. Other well-known sources of UFO reports are research balloons, as I know from personal experience. When I was a graduate student at the University of Bristol, in England, we used large polyethylene balloons to carry our cosmic-ray detectors up to altitudes of around 20 miles. A good site for balloon launching was the Royal Air Force base at Cardington, 45 miles northwest of London. This base had a gigantic hangar in which the airships of the 1930s had been garaged, and it also had a plentiful supply of hydrogen for the balloons. (Helium was too expensive to be considered in Europe, so the more risky hydrogen was used, with no mishaps for our research balloons but with fatal results for the *Hindenburg* zeppelin.) It was just our luck, with a launch in July 1954, that the winds took our balloon over London at midmorning. As the *Times* reported next day, this resulted in "thousands of telephone calls jamming the switchboards in the Air Ministry, the Meteorological Office and the police." Fighter planes were sent up to investigate and if necessary shoot down the intruder, which was reported (in the less restrained press) to be cigar shaped with flaming exhausts. Fortunately, the balloon and our equipment were then at 70,000 ft and well above the air defense's range, though not beyond the reach of imagination. Sunlight reflecting off the balloon fabric can produce a shimmering effect. When the cause of the uproar was traced to our laboratory and Professor Cecil Powell was asked about the fate of the balloon, he replied that after the payload is released to descend by parachute, the plastic "drifts on and becomes flying saucers over half a dozen different countries."[40] (A postscript: Our balloon came down across a railway line, and the laboratory got a bill for the removal of the plastic.)

Other UFOs can be traced to meteorological conditions, meteors, rocket launches, earth satellites, and of course hoaxes. In many cases, however, there is just not sufficient reliable information to be able to identify, confirm, or refute the report. In the end there is a residue of reports that are perplexing but far short of compelling any resort to extraterrestrial or paranormal theories.

Official notice was first formalized in the U.S. Air Force's Project Sign in 1947, which gave way to Project Grudge in 1949, which in turn was replaced by Project Blue Book in 1952. Many of the reports of flying objects were investigated and classified as top secret. Nothing definitive turned up to confirm either the reality of intrusions or the occurrence of novel phenomena. The air force's involvement was based on its defense role; there were concerns that perhaps some of the sightings were of advanced Soviet aircraft, in which case the United States should certainly be alerted. (One must recall that this was taking place during the cold war and McCarthy years, when anti-Soviet sensitivity was often at hysterical levels.)

For many years from the early 1950s, J. Allen Hynek, an astronomer at Northwestern University, served as a scientific consultant to the air force. His initial approach, as he has described it, was to join his colleagues in a "hearty guffaw," but over the years his view changed, and he became a believer—at least in the possibility of novel and genuine phenomena underlying some of the reports: "When the long awaited solution to the UFO problem comes, I believe that it will prove to be not merely the next small step in the march of science but a mighty and totally unexpected quantum jump."[41] Other scientists also serving as consultants to the military have been very clear in challenging the need to invoke causes outside of known science. Harvard astronomer Donald Menzel began his studies while in the U.S. Navy during World War II and has documented his observations and explanations, including "anomalous propagation" of radio.[42]

Independent and skeptical investigators also began to look into the UFO phenomenon; the writings of James Oberg and Philip Klass contain the most thorough examinations. UFO-watching clubs were formed. The United States seems to be particularly prone to theories of conspiracy and cover-up; at least with the freedom the media and authors enjoy, there is always publicity for yet another accusation that someone (usually in a government office in Washington) is trying to conceal something. And so yet another official investigation was launched in 1966, to try to deal definitively with these accusations. The U.S. Air Force commissioned a study, this time awarding a contract to the University of Colorado and putting a first-rate scientist, Edward U. Condon, in charge. The Condon "committee" (as it has been called, though perhaps it is more accurately described as a panel of consultants working with Condon and a small staff) released its report, nearly 1,000 pages long, in 1969.[43] Their failure to concede any reliable confirmation of the extraterrestrial or otherwise abnormal nature of UFOs was not, of course, accepted by UFO enthusiasts, and the report drew fire.[44] Accusations of cover-up persisted, but no official investigation has been carried out since. Nevertheless, UFO watches continue, some in a genuine spirit of research, others by people fully convinced of what they think they have seen and seeking only further examples.

The Condon committee reexamined older reports as well as those that turned up during their studies. They did more than simply interview participants, look through the Blue Book records, and study the old photographs. In two cases they had been given objects that allegedly came from spacecraft, and they subjected these to chemical and metallurgical analysis.[45] They received samples from a pile of metallic ribbon that had been found near the home of a witness who saw two spaceships overhead in 1957. The material was reported to have been radioactive when found, but nearly ten years later there was no trace of radioactivity. Analysis showed the material to be radar chaff, aluminum foil with a coating of lead powder, made by Revere Copper and Brass, Inc. Chaff is used for radar tests, as it is a good reflector of radar beams. It is taken to high altitudes by rockets and balloons and then released, to be tracked as it falls.

Condon's group also examined a metal object that was reported to have come from an airborne vehicle that exploded over Brazil, also in 1957. Through the cooperation of Jim and Coral Lorenzen, cofounders of the Aerial Phenomena Research Organization (APRO), the group obtained a sample that had been analyzed in a Brazilian government laboratory and that was supposedly of such great purity that it could not have been refined to this degree on earth through any process known at that time.[46] Condon's scientists had it checked using neutron-activation analysis, which showed the Brazilian claim to be false. The metal was mostly magnesium but with significant traces of strontium and barium and of a purity far below what was available even as early as 1940. The proportions of two magnesium isotopes showed it to be indistinguishable from normal (earth-based) magnesium. There was absolutely no evidence that this object had an extraterrestrial origin (though that has not stopped UFO groups from continuing to claim one.) The story of how the object had been obtained was so vague as to be impossible to check.

It would be nice to have at least one solid piece of any of the many UFOs that have been seen or "boarded." No sample has yet passed careful scrutiny. Reports of mere sightings are harder to check out, but at least in some cases this could be done. Investigators have been able to identify the causes of many sightings, definitively in some cases, probably in others. One well-documented set of sightings can be positively identified with the reentry into the atmosphere of the Soviet spaceprobe *Zond 4* in 1968.[47] As usually happens during such a reentry, the heat of the atmospheric friction caused the vehicle to break up, its fragments continuing forward at high speed. Each of these pieces appeared as a glowing object as it moved through the air. Larger space junk items (old satellites and rocket parts) are continually tracked by the North American Air Defense Command's radar, and the reentry of *Zond 4* was among the many that have been recorded. (A 1990 catalog from NASA lists over 10,000 objects in orbit around the earth.) Another well-documented UFO report from Tucson, Arizona, could be traced to the launch of a Titan rocket from Vandenberg Air Force Base in California, more than 500 miles away.[48]

The Condon report was clear on the central issue: whether any of the reports demanded explanations outside of the known areas and laws of science: "Careful consideration of the record ... leads us to conclude that further extensive study of UFOs probably cannot be justified in the expectation that science will be advanced thereby."[49] The report did point out that "there are important areas of atmospheric optics, including radio wave propagation, and of atmospheric electricity in which our present knowledge is quite incomplete."[50] The committee stated that they had not found any evidence that the UFOs "posed a defense hazard," and they expressed great concern at the use of UFO material in school science classes.[51]

In 1969, later in the year that the Condon report was released, the American Association for the Advancement of Science organized a symposium on UFOs. The proposal for holding this symposium had been made much earlier, but there was considerable delay, in part because of the opposition of some scientists. The speakers at the symposium were chosen to represent believers and critics. Hynek[52] and James E. McDonald of the University of Arizona were strongly critical of the official investigations, including that of the Condon committee, while Menzel was trenchant in his debunking. Papers presented at this meeting have been published (some in revised form) as a collection edited by Carl Sagan and Thornton Page, which provides a very good introduction to the study of UFOs.[53]

Some investigators have noted the psychological instability or fragility of many of the people who have reported being abducted by UFOs. Against this, a recent study failed to find any significant psychopathological difference between people who have reported strong UFO experiences (such as abduction) and people in control groups. A group of researchers in the Department of Psychology at Carleton College in Ottawa interviewed forty-nine individuals who had responded to advertisements in several newspapers. These subjects were "solid representatives of the North American middle class," most "employed in white collar occupations"; in their conclusions the Carleton group commented that their findings "clearly contradict the hypothesis that UFO reports ... occur primarily in individuals who are highly fantasy prone, given to paranormal beliefs or unusually suggestible."[54]

Resistance to the debunking continues. At the crazy extreme are people who believe that Earth is hollow and houses an advanced civilization whose vehicles (seen as UFOs) emerge through a hole at the North Pole. Adherents to this belief have charged the CIA with taking part in an ongoing cover-up of the situation. A depressingly large number of believers will never be satisfied with simple explanations, certainly not with honest expressions of uncertainty. Once we get into fantasies about visits by extraterrestrials, the literature becomes even more colorful.

Beyond the reports by people who claim to have been abducted by aliens, there was the brazen fabrication of stories attributing markings in the Nazca Desert in Peru to the landing craft of beings from outer space.

Further up the spectrum are more critical UFO believers. Books on UFOs appear regularly, demonstrating the tenacity with which this subject is still pursued and the need for great caution before accepting as fact the repeated assertions of discoveries and the accusations of cover-ups. For example, reviewing *Out There: The Government's Secret Quest for Extraterrestrials* by Howard Blum, Philip Klass documented significant errors in statement after statement.[55] Klass's review was itself attacked, with the usual allegations of bias.[56] Some seem to consider a skeptical approach a certain sign of bias (equated to a closed mind), and the only defense against this accusation would require an evenhanded presentation, as though both sides had equal merit. (This is the sort of tactic used by creation scientists; see Chapter 11.) What these critics seem not to understand is that one can be open-minded yet have a prejudice against uncritical reports or shoddy "science."

Conclusion

Science has not been a beneficiary of the subjects of this chapter. There are good scientific explanations for many of the observations and a residue of cases that have neither been explained nor yet compel us to believe that new phenomena have been seen. Much of this sort of fringe science tends to be episodic, its memory soon fading and its brashly promoted book quickly going out of print, to be replaced by yet another fantastic claim. A few topics, though, retain their public appeal. In the 1993 edition of *Books in Print*, there were eighty-three books on UFOs.

For the most part, scientists are not rushing to explore the claims. No organization of UFO believers or investigators has requested affiliation with the AAAS as have proponents of parapsychology (discussed in the next chapter). Yet the public *does* see substance behind the reports and continually confuses their "explanations" with genuine science. While a few scientists also cling to the belief of an underlying reality to some of the reports, CSICOP and other groups of skeptics continue with their efforts to debunk and demystify the UFO and other phenomena. Klass, one of the most careful of the UFO debunkers, has made an offer to "buy back every copy of [my] book (*UFOs Explained*) if at any time an authentic extraterrestrial spaceship is found."[57]

So why take time and devote so much space to these subjects? I have done this because they are prototypical of popular pseudoscience. They are immediately separable from the cold fusions and polywaters and other cases by their usually nonscientific presentations. Unreliable accounts are accumulated or invented and fancifully extrapolated into scenarios that are so implausible as to make no impact on scientists. Scientists do not rush to investigate UFOs, astrological claims, or apocalyptic earthquake predictions as they do to study cold fusion, polywater, or fifth forces. But these widely publicized works *are* considered scientific by many of their readers. One cannot expect most readers to have the time, inclination, technical knowledge, or library access to check the claims. And therein lies the problem. In the absence of (and sometimes in spite of) contrary information and a healthy skepticism that rests on some understanding of the ways of science, the nonexpert is quite likely to accept as fact this artfully promoted nonsense. This may be amusing as far as it concerns tabloid science, but the same attitude carries over to more important topics.

Notes

1. *Nature* 318 (1985): 419.
2. Ibid., 425.
3. *Skeptical Inquirer* 13, 1 (1988): 10.
4. *The Humanist*, September-October 1975, 7.
5. Later reprints of the statement included additional names.
6. *The Humanist*, September-October 1975, 7.
7. *Science in a Free Society* (London: LNB, 1978), quoted in Patrick Grim, ed., *Philosophy of Science and the Occult*, 2d ed. (Albany: State University of New York Press, 1990).
8. Michel Gauquelin, *Birthtimes, a Scientific Investigation of the Secrets of Astrology* (New York: Hill and Wang, 1983), 21.
9. *Skeptical Inquirer* 4, 2 (1979-1980): 19.
10. *The Humanist*, November-December 1975, 2; January-February 1976, 2, 28; *Skeptical Inquirer* 4, 2 (1979-1980): 19.
11. *Skeptical Inquirer* 16, 2 (1992): 150.
12. *Isis* 41 (1951): 374.
13. *Isis* 42 (1951): 3.
14. *The Jupiter Effect* (New York: Walker, 1974), 148.
15. Ibid., 89.
16. Ibid., 121.
17. *Nature* 240 (1972): 91.
18. *Science* 192 (1976): 1189.
19. *Jupiter Effect*, 127.
20. *American Scientist* 62 (November-December 1974): 721.
21. Ibid.
22. *Icarus* 26 (1975): 257.
23. Gribbin and Plagemann came out with *The Jupiter Effect Reconsidered* (New York: Vintage, 1982) seven years after Meeus wrote his piece. See Martin Gardner's review in *Discovery*, July 1982, 365.
24. *Science* 250 (1990): 511.
25. *American Men and Women of Science*, 12th ed. (New York: Jaques Cattell Press/ R. R. Bowker, 1971).
26. Browning's newsletter went to "consulting clients and subscribers," according to the St. Louis *Post-Dispatch*, July 19, 1991, 1.

27. Ad Hoc Working Group, "Evaluation of the December 2-3, 1990 New Madrid Seismic Zone Prediction," October 1990, 12.

28. G. P. Tamrazyan, *Icarus* 9 (1968): 574.

29. L. Knopoff, *The Moon* 2 (1970): 140.

30. "Evaluation," 8.

31. Ibid., 30.

32. Ibid., 31.

33. "Earthquake Preparedness Guide," October 28, 1990, 2.

34. St. Louis *Post-Dispatch,* December 3, 1990.

35. National Academy of Sciences/National Research Council, *Earthquake Prediction and Public Policy* (Washington, D.C.: National Academy of Sciences/National Research Council, 1975). For a recent appraisal, see B. A. Bolt, "Balance of Risks and Benefits in Preparation for Earthquakes," *Science* 251 (1991): 169.

36. W. Spence, R. B. Hermann, A. C. Johnston, and G. Reagor, *Responses to Iben Browning's Prediction of a 1990 New Madrid, Missouri Earthquake,* USGS Circular 1083 (Washington, D.C.: U.S. Government Printing Office, 1993).

37. "Evaluation," appendix.

38. Robert Buckhout, *Scientific American,* 231 (June 1974): 23.

39. See, for example, Margaret Sachs, ed., *The UFO Encyclopedia* (New York: Putnam's, 1980).

40. *The Times,* July 8, 1954.

41. *The UFO Experience* (Chicago: Henry Regnery Company, 1972).

42. *Elementary Manual of Radio Propagation* (Englewood Cliffs, N.J.: Prentice-Hall, 1948), and *Flying Saucers* (Cambridge: Cambridge University Press, 1953).

43. *Scientific Study of Unidentified Flying Objects* (New York: Bantam Books, 1969). Hereafter, cited as the Condon report.

44. See *Science* 161 (1968): 339, and 163 (1969): 260; also *Skeptical Inquirer* 10, 4 (1986): 328.

45. Condon report, 256.

46. Ibid., 94; Terrence Hines, *Pseudoscience and the Paranormal* (Buffalo, N.Y.: Prometheus Books, 1988).

47. Condon report, 261.

48. Philip J. Klass in George O. Abbell and Barry Singer, eds., *Science and the Paranormal* (New York: Charles Scribner's Sons, 1981).

49. Condon report, 1.

50. Ibid., 3.

51. Ibid., 5.

52. See also *The UFO Experience.*

53. *UFOs: A Scientific Debate* (New York: W. W. Norton, 1974).

54. *Journal of Abnormal Psychology* 102 (1993): 624.

55. Howard Blum, *Out There: The Government's Secret Quest for Extraterrestrials* (New York: Simon and Schuster, 1990); reviewed in *Scientific American,* February 1991, 140.

56. *Scientific American,* August 1991, 10.

57. P. J. Klass, *UFOs Explained* (New York: Random House, 1975).

9

Pscience

The general label *paranormal* covers a vast range of phenomena. An idea of their extent and variety can be gained from a listing of some of the chapters in Terrence Hines's *Pseudoscience and the Paranormal.*[1] In addition to many of the topics I have already described (for example, Velikovsky, polywater, astrology, UFOs), Hines covered psychic phenomena, life after death, laboratory parapsychology, psychoanalysis, faith healing, and health quackery, with variations under each heading.

The phenomena that are the subject of this chapter constitute what the parapsychologists call psi, including extrasensory perception (ESP) and psychokinesis (PK). ESP covers three main areas: clairvoyance, telepathy, and precognition. Telepathy is the paranormal power some people claim to use to communicate by means not otherwise known; clairvoyance is the ability of a single person to receive information, such as knowledge of distant or hidden objects; precognition is the power to foretell the future; and PK involves the movement or deformation of objects without direct physical contact. Based on intensely personal experiences, many reports contain claims for the demonstration of paranormal occurrences or of the intervention of hitherto unrecognized forces or enhanced personal abilities. These reports have been the objects of extensive scrutiny with (it is often claimed) confirmation. In ways somewhat similar to the situation with UFO reports, these claims have been vigorously contested to the satisfaction of one side or the other but not of both. Each year reports of these phenomena, well worn and new, are described and redescribed in a never-ending stream of books. Hines provides thirty pages of references after over 300 pages of text. Your local library probably has shelves devoted to books on the paranormal; in the typical shopping-mall bookstore, there is probably a section devoted to "New Age" (which may well have some other generic label by the time you read this).

Underlying all of these is the claim or belief that there are forces that humans can experience or exert, with varying degrees of conscious control, that are not part of the already well-known forces of nature. Some of the investigators believe that confirmation of their research results might well require a revision of our knowledge of the behavior of forces or of the transmission of energy. For example, the effects of all known forces diminish rapidly over distance, but a striking feature of the strength of many of the psi phenomena is their apparent indifference to distance. There is a sharp division between the firm believers in the reality of the claimed phenomena and those who deny it totally. (In the trade these are often known as sheep and goats, respectively.) In between is a relatively small group willing, in varying degrees, to examine the claims and undertake tests. But this third group seems itself to be divided. One subgroup is very skeptical but willing to investigate while applying strict scientific controls in their tests. They remain open-minded but to date have come up empty-handed. Their mirror-image subgroup has increasingly made use of modern scientific techniques but seems more sympathetically disposed to accept paranormal phenomena, more charitable toward the subjects of their tests, and more tolerant of their failures.[2]

There was no single defining episode as a starting point for psi studies. With polywater we had the papers of Fedyakin and Deryagin; with Velikovsky public awareness is well dated to the appearance of *Worlds in Collision;* with cold fusion we had the press conference in Utah. But ESP claims have been known for ages, and the literature is diffuse. The modern phase has its roots in nineteenth-century spiritualism but has since become professionalized in response to criticism both sympathetic and hostile.

In its attempts to demonstrate the reality of the phenomena, the parapsychological community organized itself as long ago as 1882, when the Society for Psychical Research (SPR) was formed in England. A similar society was formed in the United States soon afterward. One of the founding members of the British society was Lord Rayleigh, one of the giants of nineteenth-century physics. His 1919 presidential address shows a very cautious approach, stressing the need for further study. The Parapsychological Association (PA) was formed in the United States in 1957 with the purpose of promoting improved standards in parapsychological research. Each of these societies publishes a professional journal, using referees as do the better-accepted disciplines.

In 1969 the AAAS accepted the Parapsychological Association into membership in the same way that it allowed mainstream professional societies to affiliate. This action was not universally popular. In 1979 John Wheeler suggested that the PA membership be revoked. Wheeler, a physicist long respected for his many contributions to nuclear and relativity theory, noted that even ten years after affiliation parapsychology had yet to produce any scientific results. What, Wheeler asked, was a science without results? The PA, however, remains a member of the AAAS.

The earliest center devoted to psi studies was the Parapsychological Laboratory set up at Duke University in 1927 by William McDougal and J. B. Rhine. Though trained as a botanist, Rhine did much to establish the scientific approach in parapsychological research. Rhine's 1934 book *Extrasensory Perception* is a milestone in psi research. The analysis of his test data stimulated the interest of statisticians, and in 1937 the Institute of Mathematical Statistics announced its agreement with the appropriateness of the statistical methods that Rhine had used. Rhine also looked for but found no difference in the strength of psi effects when carried out over a distance as great as 250 miles.

Psi differs from the subjects I treated in earlier chapters in an important way: The phenomena it encompasses often seem to be similar to (even if more intense than) experiences of a great many ordinary people. In contrast, most people do not have personal experiences of nuclear fusion or colliding comets. In psi there is a degree of personal identification that is absent from all other pseudoscientific phenomena, perhaps excepting UFO abductions. For example, many people have had intensely vivid impressions of friends or relatives who were either distant or long dead. Most of us have had the experience of thinking of someone and soon thereafter receiving a phone call or letter from or about that person. We sometimes anticipate what someone else will say. Some coincidences seem too strange to be truly unrelated, to be the results of chance alone.

This sort of coincidence is probably the closest that most of us come to a paranormal experience, but a reasonable explanation can account for it. It is often claimed that the chance of some remarkable coincidence is only one in a billion or a trillion or some other large number and therefore one must look for less conventional causes. The weakness of this argument lies in the treacherous nature of the statistical calculation. First, it may well be true that there is only a one-in-a-million chance, but this is also true of success in a lottery, yet someone in the end does win, no matter how remote the individual chance. Second, where probabilities are small but the *number* of tests is large, the actual *number* of coincidences may not be small. This was best described by the physicist Luis Alvarez.[3] He showed how one can make a simple calculation of the chance of a purely coincidental recollection of a friend within a few minutes of learning of that person's death. Making very reasonable assumptions and pointing out that there are over 100 million adults in the United States, Alvarez calculated that nationwide one would expect there to be about ten such "remarkable" coincidences *every day*. This is not a line of argument that is going to persuade people who have no background in statistics, but it does point to the need to ask critical questions. Can there be a normal though unusual explanation? How well have the statistical odds been calculated? In fact, are the circumstances sufficiently well defined for it even to be possible to make a correct calculation of the probability? At the same time, modern parapsychologists have become more sophisticated in their experimenting, but most people are not aware of these methodological improvements. Personal experience, so widely encountered and so naively interpreted, provides a reason for the continuing public allure of part of ESP studies.

Sometimes the course of cataclysmic events provides a fertile basis for widespread personal reactions. This has been noted as a cause underlying the surge of interest in spiritualism after World War I, when many bereaved families groped for means of communicating with those who had lost their lives. What is notable is the absence of any similar effect after World War II, despite equally large losses of life. Indeed, as far as I know, there has been no widespread interest in this suggested means of communication in the Jewish communities, despite the terrible experience of the Holocaust.

How, then, does one test a person for enhanced powers? Rhine devised a special set of cards to be used in tests where the subject was asked to guess the identification of a card selected at random and placed face down. Success at rates statistically well above chance levels was reported. In a different test of psi, termed remote viewing, a target scene or object is viewed by one person, and the remote and isolated subject then describes the target. For precognition, tests are run in which the subject predicts a number that is being generated by a computer or, in PK, attempts to influence a coming number to be higher or lower. A feature of many ESP experiments has been an initial score well above that expected by chance, followed by a decline as testing continues. Many of the modern tests, such as those by Robert Jahn and Charles Honorton, to be described later, use electronic devices so that subjects obtain no sensory clues from the test apparatus and the selection of targets follows a random pattern based on a computer or the detection of the random arrival of cosmic-ray particles.

In a celebrated test of remote viewing, unusual in that it was reported in a refereed paper in *Nature*, Russell Targ and Harold Puthoff, two scientists then at the Stanford Research Institute (SRI), tested Uri Geller's clairvoyant ability to describe distant scenes.[4] The investigating scientists went to test sites hundreds of miles from Geller. At a predetermined time Geller was asked to sketch or describe the surroundings of the distant observer, whose location had not been revealed to him. Geller produced sketches that, it was claimed, were close to correct more often than not. Puthoff and Targ devised a system to score Geller's descriptions. These identifications have been challenged and the experimental procedures also criticized.

These and many more tests, demonstrations, and anecdotes have as yet yielded no positive result of remote viewing that skeptics have accepted, and there are innumerable instances where it is impossible to reach a definitive conclusion. Critics who have analyzed experiments ranging from Rhine's earliest work through today's psi studies using more sophisticated experimental regimes repeatedly raise the same issues: the possibility of cheating, either by the subject alone or by the subject and experimenter together; the sensing of cues given unwittingly by the experimenter; systematic departures from randomness in the selection of targets; inaccuracy of recording data; and disagreement over the statistical analysis. Close scrutiny by unsympathetic observers has in turn often been blamed as a cause of failure to achieve or sustain significant scores.

Before a valid theory can be constructed to draw together and provide an understanding of a range of these phenomena, if real, there must be a general acceptance of the reality of the effects, preferably extending beyond the parapsychological community. To progress beyond the hunting-gathering stage requires repeatable experiments and sufficient awareness to be able to reduce the number of variables and then concentrate on the essentials. This has proven to be extremely difficult in psi research, just as was historically the case with physics and chemistry, where progress could not be made until basic and controlled experimenting had been achieved. The biological sciences made the transition later, and in the social and behavioral sciences it is often still a problem. It is far worse in psi for two reasons. First, there is the intrusion of unconscious deception or even outright fraud. Deception is a danger that also lurks in regular science, and we always have to be watchful against the temptation to trim our results, "improve" them, or discard wayward data. Fraud is a very different matter, and I devote the next chapter to a discussion of this unhappy distortion of mainstream science. In my research I normally do not have to guard against someone's adjusting my apparatus or sneaking more cosmic rays through the counters, but the history of psi is littered with discredited experiments of earnest investigators exploited by ruthless opportunists whose skills as magicians passed for paranormal powers. In his book *Flim-Flam!* James Randi has described, in often hilarious detail, the failed attempts of various psychics to sneak something past him.[5]

A contested case in point is that of Geller, who has given numberless demonstrations of his ability to bend metal objects through the remote exercise of his remarkable powers under apparently controlled conditions. Geller was able to satisfy some scientists as to the reality of his powers, but a number of magicians, including Randi, have observed him closely and consider him to be nothing more than a very skilled magician. Randi claims to be able to reproduce all of Geller's demonstrations and to have observed Geller cheating on at least some occasions. In return, Geller has sued Randi for defamation but not yet succeeded in winning any judgment.

The most damaging exposures, however, did not involve performers such as Geller or less talented amateurs but rather implicated two researchers respected within the parapsychological community. Dr. S. G. Soal in England had studied the ability of subjects to identify cards that were drawn at random from a deck. His results showed success rates significantly better than chance. Unfortunately, in 1960 it was found that Soal had adjusted some of his test results.[6] In 1974, Dr. Walter J. Levy, who had followed Rhine as director of the Institute for Parapsychology at Duke University, was observed by two members of his research group to have tampered with his research data. He admitted it and resigned.[7]

An equally serious problem may be unsolvable within the boundaries of accepted science. It has often been asserted in psi studies that the attitude of the experimenter or observer can so affect the paranormal abilities of the subject as to make success rare or impossible. Under different circumstances this effect is well established in physics. As a component of quantum mechanics, the Heisenberg uncertainty principle shows how at the atomic level the very act of measurement influences the result. In some types of observations, it is impossible to measure simultaneously certain quantities to better than a limited precision. This limitation is far below any detectable level on the scale of everyday phenomena, and it hinges on the energy and momentum involved in making an actual measurement. It does not include the attitude or sympathy of the experimenter. But in spite of this uncertainty when dealing with an individual atomic particle, we are able to make exceptionally accurate predictions for the averages of many measurements. Nevertheless, there are allusions to the Heisenberg principle and quantum mechanics in attempts to explain the failure of various paranormal tests. To me, these excuses seem like displays of smoke and mirrors. But the point remains: If these paranormal phenomena and abilities are so sensitive that an unsympathetic investigator can induce failures or prevent successes, then we have a basic denial of the usually important requirement for reproducibility, and we would require a reformulation to include the role and attitude of the experimenter. It has been suggested that our definition of science might have to be changed to cope with these new phenomena. Most scientists are unwilling to do this—yet. Should the alleged paranormal effects come to be generally confirmed and should a change in our definition of *science* be the only way to achieve acceptance, then we would indeed have a major methodological problem.

Serious parapsychologists have long recognized the problems of fraud and inadvertent cues and have moved to meet them by trying to adopt stricter procedures, building in more precautions against being deceived. There remain differences between modern parapsychologists and their critics, such as Randi, in their estimates of the effectiveness of the precautions and the care actually taken and in their evaluation of the significance of their results.

Individual scientists, sometimes but not always affiliated with the SPR or the PA, have conducted their own investigations into these tantalizing reports of paranormal occurrences because they believe that there are surely too many and persistent reports for all to be bogus. (One response is, "Where there's smoke, there's smoke.") I have already mentioned Puthoff and Targ of SRI. John G. Taylor, a theoretical physicist, looked into Geller's abilities and was convinced; the result was his 1975 book *Superminds*.[8] Following Randi's exposure of Geller and his own further studies, Taylor completely reversed his opinion, as he described in *Science and the Supernatural,* concluding that "ESP is dead. ... I started my investigation with an open mind; the scales were not loaded on behalf of science. On the evidence presented in this book, science has won."[9]

The experiments carried out by Honorton at the Maimonides Medical Center in New York and at the Psychophysical Research Laboratories in New Jersey and by Jahn and his collaborators at the Princeton Engineering Anomalies Research Laboratory stand out for their large statistical bases, their claims for achieving highly significant success rates (statistically). Jahn, for many years dean of the School of Engineering and Applied Science at Princeton University, designed and used electronic apparatus for his large-scale testing of abilities to influence the result of the automated and random selection of numbers.[10] Subjects were asked to try to make the selected numbers increase or decrease. Jahn produced results that are puzzling and not universally accepted. The effects are small in absolute terms but seem to have statistical significance because of the large number of trials, though no single subject demonstrated any dramatic ability. For example, in 55,000 trials where the expected average should be 100.00, one subject was able to average 100.082 when trying to increase the score and 99.986 when trying to get a lower score. To explain his observations, Jahn has put forward a theoretical model based on quantum theory.[11]

As is usually the case, the critics have examined Jahn's experimental protocols and statistical analyses in minute detail and expressed their reservations.[12] Among the critics has been Jessica Utts, statistical editor of the *Journal of the Society for Psychical Research*.[13] In strictly statistical terms results such as Jahn's may seem significant *if free from systematic bias*. Small systematic effects can lead to the observed deviations from the control experiment values; it is often extremely difficult to locate and eliminate such systematic effects. Jahn has also conducted remote-viewing experiments, and these, too, have been criticized in sufficiently robust terms ("some of the poorest quality ESP experiments published in many years")[14] that they have drawn a corresponding response ("polemical generalizations and ad hominem assaults").[15]

One of my departmental colleagues, Peter Phillips, with financial support from the late James S. McDonnell (of McDonnell-Douglas Aircraft), also carried out psi studies for a number of years but found himself the object of a sting operation organized by Randi. Two young men who seemed to possess paranormal powers were able to deceive one of Phillips's associates, even though his team had been instructed in Randi's protocols to detect or avoid cheating. When suspicions arose and the procedures were tightened, the effects went away, and Randi made public his scam as a warning to overly trusting scientists.

It is legitimate to ask about these scientists who have ventured into psi research. As with the scientists involved in the cold fusion and polywater cases, they were neither cranks nor pseudoscientists. Phillips is a knowledgeable physicist. Taylor, well known for his many books that have brought current physics to the public, holds a chair at King's College in London. Jahn is an engineering professor at one of our finest universities. We need to recognize, I think, that to these and some other scientists, the paranormal initially seems to hold out the prospect of being a solvable problem, with the possibility of using modern laboratory methods to establish a major new field (as cold fusion and polywater also at first appeared—with a similar lack of success).

It takes professional courage and not necessarily misjudgment to get into this research, though most of us consider the paranormal an empty field. But it is one of the aspects of tenure and academic freedom that tenured faculty should be free to choose their research topics (within reasonable constraints such as lab space) no matter how unpopular that research might be. Having said this, I would be less than honest if I did not go on to point out that it is highly unlikely that a junior faculty member would gain tenure through paranormal research. What counts when being reviewed for tenure are results and not courageous attempts, but this is just as true with young scientists whose work is totally within the confines of traditional science.

A firm believer in at least the possibility of the reality of the paranormal was the noted author Arthur Koestler. He bequeathed over half a million dollars for the establishment in Great Britain of a chair in parapsychology. Universities did not exactly form a long line to apply for this money, and in the end it went to the University of Edinburgh, where the first incumbent of the chair has been Robert L. Morris, formerly of Syracuse University. Commenting on the establishment of the chair, the *New Scientist* expressed the widely held view that "parapsychology is still a collection of half-baked explanations in search of phenomena to explain. ... Until there is a reliable corpus of results, most scientists will continue to treat the field as an embarrassment and a joke."[16]

Morris has been concerned to establish a research program that uses careful procedures and strict safeguards. In an interview also published in the *New Scientist,* he described himself as "an agnostic, somewhat biased towards the position that new principles of nature appear to be at work here"; he cautiously noted that "to evaluate the likelihood that new means of influence exist, we must make sure we thoroughly understand the old means."[17] To date, no breakthrough has been achieved, or at least none notable enough to warrant general publicity. What has emerged from Morris and three colleagues has been a 400-page volume containing a broad review of methods and developments in the various subfields.[18] This is a sober text that discreetly avoids reference to the many harshly critical books and papers that I have cited. There is a brief mention of Randi, none of Taylor or Phillips. There are a bare two pages on "strong counteradvocacy," most of which is dismissed for its habit of focusing "on a very few media-attractive studies and essentially ignor[ing] the bulk of the research, especially those sets of experimental studies that do show partial replicability."[19] This comment displays the same error that the counteradvocates are accused of, for while much of the counteradvocacy is shrill, much is temperate and carefully documented, coming from people well trained in psychological and parapsychological experimenting.

After some years, Morris was joined by Honorton, a leading developer of the ganzfeld technique, its name taken from the German for "whole field." By this method each of the subject's sensory inputs (the "whole field") is blocked—the eyes covered and tapes of white noise played into earphones. In this isolated state the subject attempts to receive information from a tester looking at a sequence of pictures in an adjacent room. In tens of thousands of trials, a small excess (of claimed significance) of correct predictions was achieved. Honorton died in 1991, but the research program he mapped out is being continued. What is notable about Honorton's approach is that he had taken into account many of the suggestions offered by Ray Hyman, a psychologist at the University of Oregon and one of the most persistent and knowledgeable critics of parapsychological research in general. Honorton's results, showing elevated success rates above chance, have drawn typically mixed responses. Some psychologists have written them off;

others (including Hyman) have suspended judgment pending independent checks.

Another approach to the problem of validation and credibility has come from the Committee for the Scientific Investigation of Claims of the Paranormal and from the Society for Scientific Exploration (SSE). Both the CSICOP and the SSE favor critical examination of paranormal and fringe-science claims toward the end of improving our understanding of phenomena that do not appear to fit neatly within conventional science. Both adopt scientific methods. CSICOP has numerous groups of loosely affiliated skeptics around the United States who keep an eye on local pseudoscientific claims. Another organization, the Skeptical Society, publishes a peer-reviewed journal called *Skeptic*.

Parapsychology has failed to gain general scientific acceptance even for its improved methods and claimed successes, and it is still treated with a lopsided ambivalence among the scientific community. Most scientists write it off as pseudoscience unworthy of their time. A very small number, like Morris and the others I have described, think that there is enough to warrant their involvement. In a 1972 editorial called "Challenge to Scientists," *Nature* first noted the difficulty in testing paranormal phenomena and remarked that "technology has an unerring ability to suppress human skills," then went on to suggest that "a boost for parapsychological research would be very welcome. There are too many loose ends lying around for comfort."[20] Two years later, in the issue in which Targ and Puthoff's paper on remote viewing appeared, *Nature* had this to say in an editorial "Investigating the Paranormal": After detailing "highly critical comments [of the referees] that could be grounds for rejection of this paper ... it was felt that other points needed to be taken into account."[21] What clearly persuaded the editor to publish Targ and Puthoff's paper was that it was "presented as a scientific document by two qualified scientists writing from a major research establishment. ... If scientists dispute and debate the reality of extrasensory perception, then the subject is clearly a matter for scientific study and reportage."[22] In addition, "two of the referees also felt that the paper should be published because it would allow parapsychologists and all other scientists interested in researching this arguable field to gauge the quality of the Stanford research."[23]

Letters to the editor showed a widespread feeling that publication in *Nature* had conferred far too much and too premature a respectability on the parapsychological work. *Science* has not yet chosen to publish any original parapsychological papers. It did, though, publish a critical comment by George R. Price in 1955.[24] Responding to a spate of letters, some months later there was an editorial followed by short contributions from both skeptics and supporters, including Soal and Rhine.[25] More recently, the journal has expressed its generic concern about the public acceptance of the parapsychologists.[26] (This is somewhat surprising, considering that the Parapsychological Association is affiliated with the AAAS, which publishes *Science*.) At about the same time, the *New Scientist* devoted sixteen pages to Geller, with many highly critical remarks.[27] The *Scientific*

American has been notably silent on the whole issue. It does not express editorial opinions, though it has guest opinion pieces, none yet on parapsychology. In a particularly critical review in *Nature,* psychologist David Marks wrote that "the most dramatic evidence for the paranormal has been based on fraud or methodological errors."[28]

It is a political fact that sooner or later the persistence of suspicions surrounding some very public events produces an irresistible pressure for an official inquiry. The findings that emerge from these investigations are rarely accepted as definitive: Think only of the commissions that looked into the Kennedy assassination and the Iran-Contra case. Thus it was probably inevitable that reports of the paranormal would have to receive an official response of some kind. Private individuals are free to classify all sorts of items as pseudoscientific and thus beyond their obligation to dignify them with responses or research, but politicians and public agencies do not have this luxury. Whatever their own beliefs in the noncontent of the claims, it is much easier to accede to demands for a study, in the hope that an official report will be taken as authoritative and final. It never works in this way, even with scientific matters, but we have seen the same scenario played out again and again: with the AD-X2 battery additive case, with the UFO Project Blue Book and the Condon commission, and now with the paranormal.

And so in 1984, at the request of the Army Research Institute, the National Academy of Sciences set up a committee to evaluate the possible military uses of unconventional approaches to learning and training, including parapsychology. The U.S. armed forces have a continuing need to train their soldiers in the use of complex equipment; they have a mandate to ensure the security of the United States. Accordingly, as the NAS report noted, "It comes as no surprise that the Army is on the lookout for techniques that can help enhance human performance." Further,

> given the pressures and given a view of mainstream research as slow, narrow, and insufficiently targeted, it also comes as no surprise that some influential officers and certain segments of the Army want to cast a broader net to snare promising enhancement techniques. ... Proponents of such techniques are usually not content with traditional evaluation procedures or scientific standards of evidence, often giving more weight to personal experience and testimony.[29]

The committee was composed of neurologists, psychologists and people with experience in the management and training of personnel, and it examined a range of training and learning procedures for which there were claims for improvements over traditional methods. These new techniques included learning in stressful circumstances, subliminal learning, accelerated learning, and parapsychology. One committee member was Ray Hyman, already mentioned as a sharp critic of parapsychological claims. For one of its working groups, the committee solicited a paper on parapsychological techniques from James Alcock, another CSICOP member, and it also visited several laboratories, including those of

Honorton and Jahn. In 1987 the committee released its report, *Enhancing Human Performance,* which reviewed the major psi areas with possible military application: remote viewing, experiments with random number generators, and ganzfeld experiments. It also contained a good discussion of the needed scientific standards of demonstration and the methods of evaluation.

In its conclusions the group cautiously commented on a number of strategies for training, suggesting that the army be alert to techniques that showed signs of positive effects. For example, "pending further research results, the committee concludes that possible Army applications of learning during sleep deserve a second look. Findings that suggest the possibility of state-dependent learning and retention (i.e. better recall of material when learned in the same physiological and mental state) may be applicable to fatigued soldiers."[30] These topics of the committee's review do not concern us here, but we should note that the subject of parapsychology took up forty pages in the final report, and the committee's conclusions were stated in a forthright manner: "The Committee finds no scientific justification from research conducted over a period of 130 years for the existence of parapsychological phenomena. It therefore concludes that there is no reason for direct involvement of the Army at this time. We do recommend, however, that research in certain areas be monitored, including work by the Soviets and the best work in the United States."[31]

The Parapsychological Association was quick to respond with a "Special Report." The authors were highly critical of the NAS study on four major points: the concentration of the investigation on only a few parapsychological studies; the involvement of Hyman and Alcock because of their presumed prior bias; the committee's methods; and the omission of findings favorable to parapsychology. They also disagreed with many of the detailed criticisms. Their rebuttal closed with a restatement that "a strong *prima facie* case" had been made for the reality of the phenomena and with a plea for "further efforts and resources."[32] It is unclear just whose opinions have been changed by the whole exercise, but the U.S. Army can now cite the study as reason for any further inaction on its part.

Commentary

We should close by looking at a basic question: Does the lack of clear-cut results disqualify parapsychology as a science, as Wheeler and others have asserted? For comparison, we can consider a research project that (so far) has also had no positive results but continues to be studied: the search for gravitational radiation. No unambiguous signals have been detected, but there is a well-established gravitational theory from which the anticipated strength of gravitational radiation can be calculated. No detectors have yet reached the required level of sensitivity, but several antenna systems are being designed that approach the needed level. Indirect but strong evidence has been obtained by radio observations of a double star containing a pulsar. The two stars are in orbit around one another, and the rate at which the orbital speed is measured as changing is in excellent agreement with calculations that assume the system is losing energy through gravitational radiation. This research subject certainly does qualify as science, and the implications for the theory would have been severe if the observations had not brought confirmation. Official recognition of the importance of the binary pulsar came with the award of the Nobel Prize in physics in 1993 to Joseph Taylor and Russell Hulse, its discoverers. Suppose, though, that there were no guiding theory, only a suspicion that gravitational radiation should exist, and suppose that no positive results had been obtained after more than a century of searching. Would we still feel it worth continuing with the search, and would we classify the study as scientific? I think not.

Another argument often raised against research into the paranormal is that there is no satisfactory theory to explain some claimed results. This may often be true, but it is not relevant. As I have pointed out in previous chapters, the history of many genuine discoveries shows that there was no preceding theory. The discovery of radioactivity long predated any theory. Cosmic rays came out of the blue. There was no theory in which the discovery of the coelacanth filled an important gap. There *are* theories that can guide us to discoveries or provide a framework into which some discoveries can be fitted, but some discoveries can start as anomalies that—when confirmed—will require the construction of a new theory or the modification of an old one.

Where, then, does parapsychology fit in the subscientific fringe area? It has so far failed to produce any clear evidence for the existence of anomalous effects that require us to go beyond the known region of science. The best that it has achieved is, for example, a deferred judgment from a skeptical expert such as Hyman. There is no science-based expectation that definitive results will be forthcoming, unlike the case of research into gravitational radiation, where increased instrumental sensitivity must yield signals or a well-established theory will have to be critically reexamined. Even though competent scientists are engaged in parapsychological research, I am very hesitant to classify it as science, though I have no reluctance in tagging as pseudoscience the work of the magicians and others who have traded on their sleight of hand.

An interesting and different view of parapsychology has come from Trevor Pinch, a sociologist of science. He has discussed the demarcation problem and considers that fraud "is the principal normal counter-explanation for the parapsychological evidence."[33] In view of the research such as that of Jahn and Honorton, Pinch's assertion does not seem adequate. Pinch expresses the view that demarcation arguments are culturally dependent and argues "that the rejection of parapsychology rests on cultural differences which demarcation criteria serve to legitimate."[34]

And there matters rest, though not with any sense of finality. Parapsychological research will certainly continue because some people are convinced that the phenomena are real and because others, coming to this subject for the first time, are curious and think that they can improve on the methods used. Skeptical scientists, far outnumbered by the believers and the curious, will continue to view this research with disdain or worse. As is the case with UFOs, there will probably continue to be a small number of scientists who believe that there are reports that have not been satisfactorily explained within conventional science and that the need for further research remains. A gullible public will continue to be exploited by those whose interests are somewhat less than scientific. What is noteworthy, though, is the sameness of the debates over the years. In 1926 there was an extended correspondence in *Nature*, touched off by a review of Arthur Conan Doyle's *History of Spiritualism* and concluding with a very long editorial four months later. We find there comments that have been repeated whenever the subject comes up: "As the control becomes stricter the results become poorer. When the control becomes rigid the phenomena cease altogether. That is the general rule, and it admits of only one interpretation."[35] Sixty years later, Hyman, the psychologist, noted that "the level of the debate during the preceding 130 years has been an embarrassment for anyone who would like to believe that scholars and scientists adhere to standards of rationality and fair play."[36] Each generation's best case for psi is cast aside by subsequent generations of parapsychologists and is replaced by newer "best" cases. It has been, in the words of that noted philosopher of science Yogi Berra, a case of déjà vu all over again.

Notes

1. *Pseudoscience and the Paranormal* (Buffalo, N.Y.: Prometheus Books, 1988).

2. See, for example, John Beloff, *The Relentless Question: Reflections on the Paranormal* (Jefferson, N.C.: McFarland, 1991); reviewed by Martin Gardner in *Free Inquiry*, Winter 1990-1991, 55.

3. *Science* 148 (1965): 1541.

4. *Nature* 251 (1974): 602; see also 274 (1978): 680 for critical comment.

5. *Flim-Flam! The Truth About Unicorns, Parapsychology, and Other Delusions* (New York: Lippincott and Crowell, 1980).

6. *Nature* 245 (1973): 52.

7. J. B. Rhine, *Journal of Parapsychology* 39 (1975): 306; see also Patrick Grim, ed., *Philosophy of Science and the Occult*, 2d ed. (Albany: State University of New York Press, 1990), 253.

8. *Superminds* (New York: Viking, 1975).

9. *Science and the Supernatural* (New York: Dutton, 1980).

10. *Proceedings of the IEEE* 70 (February 1982): 136.

11. R. G. Jahn and B. J. Dunne, *Foundations of Physics* 16 (1986): 74.

12. See C.E.M. Hansel, *The Search for Psychic Power* (Buffalo, N.Y.: Prometheus Books, 1989), and Ray Hyman, *The Elusive Quarry* (Buffalo, N.Y.: Prometheus Books, 1989); see also Ray Hyman, *Proceedings of the IEEE* 74 (1986): 823.

13. *Statistical Science* (1991): 363; this paper is followed by over twenty pages of comments.

14. G. P. Hansen, J. Utts, and B. Marwick, *Journal of Parapsychology* 56 (1992): 97.

15. Y. H. Dobyns, B. J. Dunne, R. G. Jahn, and R. D. Nelson, *Journal of Parapsychology* 56 (1992): 115.

16. *New Scientist* 98 (June 30, 1983): 922.

17. Ibid., 107 (September 5, 1985): 59.

18. *Foundations of Parapsychology* (London: Routledge and Kegan Paul, 1986).

19. Ibid., 322.

20. *Nature* 246 (1972): 321.

21. Ibid., 251 (1974): 602.

22. Ibid., 559.

23. Ibid.

24. *Science* 122 (1955): 359.

25. Ibid., 123 (1956): 9.

26. Ibid., 184 (1974): 1233.

27. *New Scientist* 64 (1974): 170.

28. *Nature* 320 (1986): 119.

29. Daniel Dudman and John A. Sweets, eds., *Enhancing Human Performance* (Washington, D.C.: National Academy Press, 1988), viii.

30. Ibid., 19.

31. Ibid., 22.

32. John A. Palmer, Charles Honorton, and Jessica Utts, *Reply to the National Research Council Study on Parapsychology* (Research Triangle Park, N.C.: Parapsychological Association, 1988), 19.

33. *Social Studies in Science* 9 (1979): 329.

34. Ibid.

35. *Nature* 118 (1926): 721.

36. *Proceedings of the IEEE* 74 (1986): 823.

10
Fraud: Yes, No, and Maybe

In leading you through this survey around the margins of science, I have wanted to show how genuine science differs from incorrect science and also how some truly revolutionary ideas have been incorporated into science, while others that make apparently plausible claims can be promptly rejected. At the same time we should recognize how difficult it may be to make these distinctions. We have seen how honest error can come to light as scientists try to repeat experiments, how genuinely scientific and correct ideas might take a long time to gain acceptance but in the meantime are viewed as eccentric, and how pseudoscientific ideas can often be quickly and correctly so identified. We have also seen how some ideas have not yet produced any positive results but can differ widely in their plausibility and thus in their attractiveness as subjects for research. We come now to a subject that is gaining increased attention. Some seemingly genuine science, carried out within the mainstream, has turned out to be not simply wrong but fraudulent. There have also been instances where data have been carefully manipulated yet the results have been confirmed. Inevitably (and rightly or wrongly), these revelations tend to tarnish all of science, though the stains are unevenly distributed.

In order to keep this chapter to manageable proportions, I concentrate on the ways in which the "frauds" have taken place, how they have been discovered, and the remedial practices that are being implemented. Motivation for fraud gets us into a morass of speculation: Some general factors are sufficiently obvious and agreed upon, but assignment of reasons in individual cases is often so speculative as to be unscientific, and I try to avoid attributing reasons where possible.

At the start I want to exclude historical cases of fraud (alleged or proven), which I loosely define as being before about 1900. There are several main reasons for this decision. Many of the historical cases of alleged fraud or tampering that have received attention are those of famous scientists—Galileo, Newton, Mendel, for example—whose correct analyses are alleged to have come at least in part from bias or contaminated data from which discrepant results were subjectively removed. The conditions in science in the past have differed so much from those prevailing today as to make comparisons and lessons unreliable. We often have insufficient documentation. There are modern cases where fraud has been alleged or demonstrated but where other scientists have confirmed the result. I have two cases to include in this category, relating to Robert Millikan's experiments with electrons during the early years of the twentieth century and Cyril Burt's analyses of the IQ of twins. As far as I know and apart from the Baltimore case (covered later in this chapter), no current allegations of fraud involve correct results. I know of active scientists who have been wrong and defended tooth and nail what was patently incorrect, but that is a very human failing. Then there are those cases where the results have been shown to be fraudulent and wrong; we look at some examples of this type first.

In the past scientists widely believed that science possessed sufficient internal checks to effectively deter fraud, to discover dishonesty so quickly and efficiently that the resulting damage to a scientist's professional career would be too great to risk. But times have changed, as is shown by the increased incidence of proven cases of dishonesty. It will be instructive to start with a look at some representative cases. We can then try to see whether there are any overall patterns and finally consider remedies that are being introduced.

There have been many articles on various aspects of scientific fraud, but the most comprehensive and widely quoted though contentious treatment is to be found in the 1982 book by the experienced science writers William Broad and Nicholas Wade, *Betrayers of the Truth*. *Fraud* is usually taken to imply deceit, the knowing representation of something as true that is not really so, but here I want to caution that the label of fraud is not always or necessarily easy to apply.

Although I deal with this point again in discussing individual cases, I should also point out that some of the allegations of fraud have not yet been proven to the same degree required for a conviction in a court of law. Innocent reasons have been put forward to explain seemingly clear-cut instances, but once thrown, much of the mud sticks. One might expect that in science cases of fraud could be settled firmly one way or the other, but ambiguity exists for many reasons. The evidence may be incomplete because important notebooks or papers cannot be found or are known to have been destroyed, principals are dead or cannot be found, and participants' recollections are contradictory. Furthermore, even a clear case of data fabrication may have little practical importance if the major results are the same with or without the tampered data. The manipulated data can tell us about the character and working habits of the accused but may have no effect on the progress of science.

Cyril Burt and Psychological Research

There is a still running and often vehemently debated controversy surrounding the issue of "intelligence": What is it, and how can it be evaluated? Is there one type of intelligence, or are there many? To what extent can we define intelligence and measure it? Is a person's intelligence determined by heredity or environment or some combination, and, if the latter, in what proportions? Large-scale social policy has hinged on the answers to these questions, and so, inevitably, the political views and preferences (and alleged views and preferences) of the participants have added a heat and color that is fortunately missing from most scientific disputes and their pseudoscientific imitators.

The social importance of the question has been shown on both sides of the Atlantic. In England for many years, schoolchildren were subjected to a general test at age eleven. Known as the 11+ exam, the test was the basis for deciding the type of high school the child could attend. One track led to academic programs and the possibility of entry into a university, the other to "secondary modern" schools and less socially prestigious careers. The pressure on the children and their families was often intense, the results of the tests determinative of lives and accordingly embittering. In the United States the most recent and especially inflamed heredity-versus-environment fight was started by a 1969 paper by Arthur Jensen in the *Harvard Educational Review* and a 1971 article by Richard Herrnstein in the *Atlantic*. Since then, the issue has continued to smolder, spilling over into debates on educational policy and the desegregation of schools through the widespread busing of students. It has also been suggested, and disputed, that research on IQ was a motivating factor in the U.S. government's passage of restrictive legislation on immigration.

Sir Cyril Burt was one of the major figures in the development of psychology. His biographer, Leslie Hearnshaw, described Burt's pioneering role in creating the profession of psychology: "He was the first person who was primarily a psychologist ... the first Britisher to devote his life simply and solely to psychology, who was paid for being a psychologist, and who never practiced in any other field."[1] His contributions were recognized as early as 1946, when he was awarded a knighthood. When he retired from his chair at University College, London, in 1971, his position in psychology was elevated and secure.

Burt's work on quantitative psychology and intelligence testing was a major component in the decision of the British educational authorities to institute the 11+ exam after World War II. The basic assumption was that by age eleven a child's intellectual ability could be reliably measured so that the child could be directed into the education and career commensurate with those abilities—measured by what are generally termed IQ tests. The thesis accepted at that time was that heredity (genetics) played the major role in determining intelligence. Burt had been one of the pioneers in the development and use of factor analysis, the statistical technique by which the influences of several contributory factors can be separated within a composite measurement. Thus Burt's research had shown that about 75 percent of "intelligence" could be attributed to inherited ability, and lesser proportions depended on general and personal environmental influences.

Burt had reached his conclusions from the study of monozygotic (identical) twins who had been raised separately, comparing their similarities with those of twins raised together. Burt's compilation of cases contained the largest data pool then assembled, and his conclusions carried great weight, especially when considering his standing in quantitative psychology.

Shortly after Burt's death, his sister invited Hearnshaw to write a biography of Burt. Though neither a student nor a close associate of Burt, Hearnshaw held a chair of psychology in the University of Liverpool and had several years before written a history of British psychology. During the years that Hearnshaw was working on his biography of Burt, accusations began to appear that many of Burt's important results had been fabricated or at least manipulated. Leon Kamin, a psychology professor at Princeton, made these accusations during a seminar in 1972 and in his book *The Science and Politics of IQ* in 1974. Wider publicity followed in 1976 with articles by medical correspondent Oliver Gillie published in the *Times*. The changing view of Burt is reflected in Hearnshaw's description of how he "became convinced that the charges against Burt were, in their essentials, valid,"[2] and his book documents in great detail Burt's career and his now increasingly disputed research findings.

Kamin had become suspicious of some of Burt's statistical analysis, and his investigations led him to other inconsistencies and problems within Burt's work. Kamin noticed that some of Burt's numbers showed far too good an agreement. Statistical tests can measure the degree to which two quantities are related; the relationship can be expressed in terms of a correlation coefficient, which has values from 1.000 (for perfect agreement) to zero (for no agreement at all). Burt had listed 0.944 for the correlation between IQs for identical twins raised together and 0.744 for identical twins raised in different homes. These are very high values, and together they give strong support for the idea that heredity is the major factor in intelligence. What Kamin noted was that these two values remained the same (to all three decimals) even though they appeared in research papers between 1955 and 1969 and related to twenty-one twins in the earlier study and to fifty-three in the later one. To say that this sort of agreement is statistically unlikely would be a gross understatement. In defense of Burt it has been suggested that there is an innocent explanation: that Burt knew there was supporting literature from other researchers and that more data from him would not have much effect. So in his old age Burt may simply have used the old values rather than go through a tedious recalculation. Even if this surmise were correct, it should not be seen as condoning Burt's actions, and it certainly does not exculpate him from other charges.

Supporting evidence of Burt's manipulation of his data has come from, among others, Donald Dorfman, a psychologist at the University of Iowa, who concluded that "beyond any reasonable doubt, Burt fixed the row and column totals of the tables" and that the data "were fabricated."[3] More recently, a defense of Burt appeared in Robert Joynson's book.[4] This book has in turn received conflicting reviews. For example, Steve Blinkhorn was critical of Joynson's analysis of some of the mathematical arguments.[5] For his part, Ronald Fletcher was generally supportive of Burt and was inclined to lay on the media much of the blame for the acceptance of the accusations made against Burt.[6]

Gillie reported that he had not been able to track down two of Burt's coauthors, Margaret Howard and Jane Conway. There is now conflicting evidence on the reality of their existence. On the one side are allegations that they and other coauthors were simply products of Burt's imagination. On the other side John Cohen, professor of psychology at the University of Manchester, wrote the *Times* to record that he had met Howard;[7] her reality was also evident to one of the twins whom she had tested.[8] The whole affair is complicated because Burt was among the many academics evacuated from London during the war; some of his records were lost, others destroyed in bombing raids.

Many other accusations and rebuttals have been traded, accompanied by fusillades of pejoratives. I cannot evaluate the merits of the pro and con sides to select a winning side (if there is one), but I can turn our attention to my reasons for selecting this case. The IQ controversy with its attendant social applications has been a subject of disagreement for years. Why did it take so long before Burt's data were examined sufficiently closely for questions to be raised? Burt's papers were widely read and quoted and were enormously influential, but as the later detective work showed, they provided inadequate data for independent checks. There were conflicting descriptions of how the twin data had been obtained. Psychology students are now required to take courses in statistics, but at the time the Burt papers were appearing, few psychologists could match Burt's mathematical capabilities, and even fewer seemed willing to confront his known polemical skills. The overall picture one gains is of a commanding professional presence, unwilling to accept criticism and intimidating to potential critics, though there were rumblings in some quarters. In addition, for sixteen years Burt was editor of the *British Journal of Statistical Psychology*, where he exercised tight control. It seems as though his own papers did not receive the editorial scrutiny that they needed.

In 1988 Mark Snyderman and Stanley Rothman did draw attention to "some rather large quantitative discrepancies" between different studies yet noted the "quantitative agreement between genetic relatedness and IQ correlation in the kinship data."[9] For their analysis, they excluded Burt's data but added the comment that "even if [he] did fake much of his data, he was careful to manufacture reasonable figures, as the inclusion of his results make little difference to the median values calculated."[10] A similar comment was appended in a footnote to the Minnesota study of twins: The "correlation reported by the late Sir Cyril Burt and questioned for its authenticity after his death falls within the range of findings reviewed here."[11]

In 1980 the British Psychological Society lent its support to the allegations of fraud, but twelve years later the society decided to express no official position, in spite of strong pressure from a number of its senior members. While this case provides some clear lessons regarding the need for independent testing and analysis and some warnings against the unwillingness to raise questions, the picture is not really so simple. What this case does illustrate is Burt's departure from the scientific standards that we expect. It also illustrates the general habit of scientists to treat with respect work in the published literature, at least to the extent of only rarely being willing to undertake the tedious chore of recalculating someone else's analysis unless there is some strong reason to doubt the result.

N-rays

The turn of the century was an exciting time in physics. Only a few years earlier, some physicists believed that all the major discoveries had been made and all that remained was the improvement of experimental accuracy with the increase in the number of decimal places. Then came the totally unexpected discoveries of X-rays by Wilhelm Conrad Roentgen, radioactivity by Henri Becquerel and the Curies, and the electron by J. J. Thomson. In 1900 Planck introduced the first quantum ideas in his theory of heat radiation. There was, though, as yet no idea of the complexity of the structure of atoms and their nuclei. And so in a sense it was not a surprise when René Blondlot, a professor of physics at the University of Nancy, announced his discovery of yet another kind of radiation, which he named N-rays in honor of his hometown and university. Blondlot was widely respected. He had shown (though not very accurately) that the new X-rays traveled with the same speed that light did, and his work had been recognized in 1893 and 1899 by the award of prizes by the Academy of Sciences.

Blondlot continued his studies of the properties of X-rays. In his experimental setup he used the standard means of producing X-rays by the application of very high voltages to a cathode-ray tube (CRT, the forerunner of the modern TV tube). The high voltage accelerated electrons inside a sealed glass container. The electrons slammed into a metal target at high speed, and X-rays were produced in the rapid deceleration of those electrons. The X-rays sprayed in all directions, and lead sheets were usually used to confine the rays and to shield other apparatus from the radiation. The cathode-ray tube itself glowed with a green hue where X-rays hit the glass.

Blondlot ran a pair of wires from the high-voltage leads, keeping the ends of the wires a small distance apart. When the high voltage was switched on in the CRT, a spark jumped across the gap between the exposed wire ends. Blondlot began to note changes in the brightness of the sparks, and further experimenting showed how those changes depended on a number of parameters. He reported that the effects were most pronounced for very weak and short sparks—sparks less than .1 mm in length. Blondlot observed these changes when lead and aluminum screens shielded the spark gap from radiation from the CRT with its X-rays.

Blondlot considered that he must be detecting some new type of radiation, and he confirmed this when he found that his new radiation could be refracted by a quartz prism, something that X-rays did not show. The new radiation also differed from X-rays in that it did not produce fluorescence in certain materials, and neither did it affect sensitive photographic materials.

As one would anticipate, Blondlot's announcement of his results attracted considerable attention, some of it critical. Heinrich Rubens, a distinguished German physicist who had made many infrared radiation measurements, was unable to reproduce Blondlot's results, but Auguste Charpentier, a colleague of Blondlot, found N-rays to be emitted by animals and humans. Blondlot expanded his experimenting and found that N-rays could also be detected coming from a special gas-burning lamp, from the sun, and indeed from any glowing object. Throughout, the effect remained weak, requiring darkened surroundings and a considerable time for the experimenters' eyes to adapt to the dark in order to see the faint brightness changes.

While confirming reports continued to come out, so did critical comments. It was shown that some of the effects were psychophysical and could be attributed to the behavior of the rods and cones making up the retina of the eye. A visitor to Nancy reported that some experimenters did see changes in the spark intensities but others did not. Charpentier, whose observations had been early confirmations, knew of the retinal behavior; indeed, his own doctoral thesis had been concerned with the behavior of the retina. Blondlot and others, sensitive to the problem of having only visual and subjective measurements, devised photographic methods of recording the sparks. Blondlot even announced the discovery of a variant of his N-rays and named them N'.

These discoveries and the critical comments alike were discussed at scientific meetings. At the Cambridge meeting of the British Association for the Advancement of Science, Rubens suggested to the visiting American physicist Robert Wood that he visit Nancy. Wood, from Johns Hopkins University, was an expert in optics, author of the standard text *Physical Optics* that came out in 1906 and was in use for many years. In a letter to *Nature*, Wood reported on his visit to the Nancy laboratory:

> I went, I must confess, in a doubting frame of mind, but with the hope that I might be convinced of the reality of the phenomena. ... After spending three hours or more in witnessing various experiments, I am not only unable to report a single observation which appeared to indicate the existence of the rays but left with a very firm conviction that the few experimenters who have obtained positive results have been in some way deluded.
>
> I suggested that the attempt be made to announce the exact moments at which I introduced my hand into the path of the rays. ... In no case was a correct answer given. ... The fluctuations [on the screen] bore no relation whatever to its movements. ...

I was next shown the experiment of the deviation of the rays by an aluminum prism. ... I subsequently found that the removal of the prism ... did not seem to interfere in any way with the location of the ... ray bundle.[12]

Some more revealing detail is quoted in a 1941 biography of Wood: "I asked him to repeat his measurements, and reached over in the dark and lifted the aluminum prism from the spectroscope. He turned the wheel again, reading off the same numbers as before. I put the prism back before the lights were turned up."[13]

Wood's report also appeared, in translation in the *Revue Scientifique* and the *Physikalisches Zeitschrift*, but N-rays did not vanish completely and promptly. Most of the scientists who had confirmed Blondlot's findings were unpersuaded, though in 1905 only one new confirmation appeared. What is most remarkable is the persistent support for the N-rays in Nancy. Blondlot, though, took early retirement in 1909 and continued his N-ray observations for a number of years.

By now N-rays have almost been forgotten, recalled mainly in the books and articles that list the well-established cases of scientific fraud. Closer inspection and consideration of nonscientific factors suggest that the N-rays should not be quite so easily classified. The starting point was Blondlot's use of visual and necessarily subjective estimates of changes in the spark intensities. But visual estimates of brightness have been used for centuries in astronomy. Their quantitative limitations came to be recognized when photography and later electronic methods came into use in the nineteenth century, but the eye is a remarkable instrument, and experienced observers still have a place. Working close to the limits of visual sensitivity, as Blondlot was doing, only increases the chances of fooling oneself, which is surely what Blondlot did at the start.

But how do we explain his persistence and the confirmation that he received from other scientists? Persistence, even when wrong, is not uncommon. I have seen it in cosmic-ray work, where papers have appeared reporting the discoveries of new types of subatomic particles, and the authors have tenaciously held to their interpretations—even as the years passed with no confirmation, and, more tellingly, several rebuttals showed that more pedestrian interpretations were probable. Self-deception, though, is not the same as deliberate fraud. But what about Blondlot's supporters? Here we are indebted to Mary Jo Nye for her thorough analysis of the N-rays affair.[14] She put together a very plausible picture of science in general and physics more particularly in France at the turn of the century. She suggested that in essence the playing out of the N-ray case was a consequence of nationalistic feelings enhanced by a strong regionalism. The scientists who claimed to have confirmed Blondlot's results came from Nancy; the critics, from the start, were from many different laboratories in the United States, Britain, Germany, and even other regions of France.

Nye draws a picture in which the prestige of France was in eclipse. Immediately prior to the discovery of N-rays, another French physicist, Victor Cremien, had tried without success to repeat some electromagnetic experiments of the American Henry Rowland. A joint U.S.-French experiment had shown Cremien to have been wrong. Concern was being voiced in France that American graduate students were going to study in Germany rather than in France. Discovery of the N-rays seemed at first to restore the reputation of French science, but with later unfortunate results.

Are there lessons from N-rays that relate to the science of the late twentieth century? The danger of self-deception is foremost, always a potential hazard. It is all too easy to be convinced that one is correct and that others fail to understand or lack the experimental skills. Protection comes from a willingness to listen to critics, not necessarily to agree with everything but at least to learn something. There are echoes of the N-ray episode in the cold fusion case.

Medical Fraud

Burt's manipulation of his data and Blondlot's persistence in his self-deception are both rather sad examples of scientists with established reputations displaying their frailty. The next few cases I group together can perhaps be better described as sordid. They involve blatant dishonesty in the fabrication of scientific credentials and/or data or plagiarism on a breathtakingly audacious scale. All have turned up in the medical sciences. Allan Mazur has provided a comprehensive review with discussion of responses to the allegations.[15]

The case against John Darsee has been set out in detail by Walter Stewart and Ned Feder in an article in *Nature*.[16] Darsee had held positions in the medical schools at Emory University and Harvard. In 1981 some of his colleagues found that he was inventing his primary data. When confronted, he admitted this and was allowed to continue his research, only to be found to be doing it again. Investigating committees at the two medical schools showed clearly the pattern of deception, which was the subject of Stewart and Feder's analysis of 109 papers published by Darsee. Their analysis concentrated on matters of internal consistency (or lack of it) that they believed could and should have been caught by Darsee's coauthors and the editors and referees of the journals where the papers appeared. Stewart and Feder contend that scientists have been too trusting and that coauthors, editors, and referees should exercise greater responsibility to guard against manipulation of the system by aggressive and unscrupulous individuals. Their thesis has been strongly criticized in turn. In the same issue of *Nature,* there were counterarguments in an editorial as well as in a vigorous response from Eugene Braunwald of the Harvard Medical School.

Scientists at Harvard considered Darsee to be exceptionally gifted, and there was an initial reluctance to entertain the serious allegations that were made against him. It was similar to the case of Dr. Elias A. K. Alsabti, who turned out to be a confidence trickster of truly awesome proportions. Independently wealthy and claiming to be related to the Jordanian royal family, he deceived physicians at a number of U.S. hospitals and medical schools.[17] Alsabti's special talent lay simply in plagiarism. He would take published papers, retype them, substituting his name for that of the genuine authors, then submit them to obscure medical journals. Suspicion emerged when the editor of a Japanese journal showed that Alsabti had copied a 1977 article and published it under his name two years later. Alsabti's plagiarism was made public in *Nature* by Akiva Shishido of the National Institute of Health in Tokyo, and the *Journal of Medical Sciences* published a notice of retraction.[18] A number of other journals also printed retractions, but many of Alsabti's "publications" remain on the record.

M. J. Purves, a reader in the Department of Physiology at the University of Bristol, in 1981 wrote to his department, acknowledging that "data published in the proceedings of the 28th International Congress of Physiological Sciences ... are false. ... None of my colleagues were involved."[19] Purves resigned from his faculty position after an internal investigation.

Many years before this, six members of the editorial board of *Science* published a letter that stated, "We have had our attention drawn to several discrepancies in a paper ... which was recently published in *Science*." They went on to describe how one figure "was merely an enlargement of a part of [another] figure" and "tilted at a slight angle." Some of the data were not as represented; a third figure was not original (as claimed) but had been lifted from a paper of another scientist. "We apologize to our readers for this unfortunate event," wrote the editors.[20]

In 1988 Stephen Breuning, a psychologist at the University of Pittsburgh, entered a plea of guilty to a charge of falsifying data in his application for federal research funds. Breuning's research involved drugs used in the treatment of mentally retarded children. Robert Sprague of the University of Illinois, who visited Breuning's laboratory in 1983, became suspicious of Breuning's results because independent tests showed far too close an agreement. An investigation by the National Institute of Mental Health (NIMH) found that Breuning had knowingly misrepresented his results. The prosecution was initiated because of the involvement of federal funds.

The 1986 case of Claudio Milanese also hinged on fabricated data. A short letter in *Science* signed by Milanese and two colleagues at the Dana-Farber Cancer Center and the Harvard Medical School made a terse announcement that "those biological data are incorrect" and that they wished "to retract the data and the conclusions based on them." The three scientists extended their "apologies to the scientific community," trusting that "certain misinformation presented in that article can be rectified by publication of this retraction letter."[21] A news report in the same issue of *Science* told how other researchers had found that their experiments were not working after Milanese had returned to Italy. An investigation was started, and in a letter Milanese admitted that the data were not valid.

Another example of data fabrication came to light at Massachusetts General Hospital in 1979 when John Long was reluctant to show his notebooks to a colleague. When at last he did produce them, they turned out to be forged. Until that time Long's work had been well regarded, receiving major funding from the National Cancer Institute. After Long's resignation, the detective work continued, and his unusual results were traced to contamination in the human cells he had used.[22]

These examples of fraud have been in medical science, where research funding has been on a generous scale, awarded through an extremely competitive process. Scientists in other disciplines are not necessarily any less frail or open to temptation, yet geology, botany, astronomy, and the other sciences have produced nothing like this steady trickle of fraud. (In later chapters I analyze the issue of fraud, its origins and implications, but in the present chapter I simply catalog and describe the phenomena.)

Unresolved Allegations

In contrast to the preceding section, where there is no doubt regarding the fraud, the two cases that come next are sufficiently fuzzy that they are currently still the subjects of disagreement and investigation.

Himalayan Fossils

In April 1989 *Nature* began a series of reports, "The Peripatetic Fossils." The opening shot in this continuing exchange came from John Talent, a geologist at Macquarie University in New South Wales, alleging that the work of Jit Gupta of Punjab University was not valid. Talent alleged that fossils that were supposed to have been found in the Himalayas had actually come from Europe and East Africa.[23] This accusation brought a pained response from Gupta and, in the same issue, a note from one of Gupta's colleagues, A. D. Ahluwalia, saying that most of Talent's doubts about Gupta's work were "well-founded."[24]

Philippe Janvier, who had collaborated with Gupta on some papers, wrote to express gratitude to Talent for his work, and pointed both to the pressure to publish as well as difficulties in "defining what is fraud in paleontology."[25] A few months later J. B. Waterhouse of the University of Queensland, another of Gupta's collaborators, stated that he could "personally vouch that much of Gupta's material emphatically came from the Himalayas."[26] There was yet another response from Gupta, and another from Talent. In March 1990 Punjab University announced that it was planning a field expedition to visit the disputed Himalayan sites as soon as sufficient financial support had been assured.

Less than a year later, the *Journal of the Geological Society of India* reported on the results of its investigation of some two dozen of Gupta's 400 publications. The journal concluded that there was support for the accusations that had been made against Gupta, and it recommended that its readers disregard all papers of Gupta it had published between 1969 and 1988.[27]

As an editor in *Nature* noted, "The debate hinges on obscure details of Himalayan paleontology and stratigraphy, made no easier in the reading of place-names where most people never go. ... But close reading of the accusations and responses leaves the impression that Gupta's defense is flimsy."[28] In February 1992 Punjab University reinstated Gupta to its faculty but did not reappoint him as head of the department. In the welter of accusations and rebuttals, the questions remain.

The Baltimore Case

Troubling as are all of the cases of dishonesty (proven or alleged) that I have described, they are dwarfed by the next one, in its implications for the conduct both of scientific research and of investigations of allegations of academic dishonesty. It started in April 1986, when the professional journal *Cell* published a paper on some research in immunology carried out by Thereza Imanishi-Kari, Margot O'Toole, and David Baltimore.[29] The principal author was Imanishi-Kari, then at MIT and now at Tufts University; O'Toole was a postdoctoral fellow; Baltimore, a Nobel Prize winner, was then director of MIT's Whitehead Institute for Biomedical Research and later became president of Rockefeller University. Perhaps unfairly this saga has come to be termed the Baltimore case, not because Baltimore has himself been accused of dishonesty but because of his prestige and association with a colleague whose actions have been the focus of a continuing investigation of unprecedented scope.

It appears that for some time O'Toole had felt that her laboratory data had been misinterpreted by Imanishi-Kari; ultimately, the allegations prompted informal reviews at both MIT and Tufts. Both reviews found no evidence for dishonesty but rather considered O'Toole's views as simply representing a legitimate and different interpretation of the data. The disagreement, they felt, pointed to the need for more data. O'Toole persisted with her allegations, and news of these suspicions reached Walter Stewart and Nick Feder at the NIH, who have developed a reputation for investigating allegations of dishonesty in scientific research. (Stewart was a member of the team that *Nature* had sent to Benveniste's Paris laboratory and, together with Feder, had investigated the Darsee case described earlier in this chapter.) Feder and Stewart claimed that the internal reviews at MIT and Tufts were not sufficiently thorough. In January 1988 NIH sent a special investigating team to Boston, but the panel was soon reconstituted after complaints had been made that two of the panelists had professional associations with Baltimore. In February 1989 the new panel rejected charges of fraud and misrepresentation. Meanwhile, in April 1988, Congressman John Dingell had started hearings before his Subcommittee on Oversight and Investigations of the House Energy and Commerce Committee. These hearings dragged on for more than three years, and they took on a complexion that at the time many scientists regarded as sinister.

Initially, only O'Toole was called to testify and not Baltimore, Imanishi-Kari, or the members of the MIT and Tufts review panels. When Baltimore at last testified, there were heated exchanges between him and the chairman. The Secret Service was asked in to examine Imanishi-Kari's laboratory notebooks and from this emerged the further allegation (from ink and paper analyses) that various entries had been altered and had not been made on the dates claimed. Photographs of the "doctored" notebooks appeared in *Science*.[30]

This case has raised a number of very serious questions without resolving the original allegations. In the National Academy of Sciences journal *Issues in Science and Technology*, Baltimore identified the major issues: "Who should judge science? what is the worth of scientific collaborations and what are the duties and responsibilities of those involved in them? is error in science inherently bad and should it always be corrected when found? how does one distinguish between error and fraud? and does science do an adequate job of policing itself?"[31] These questions merit discussion, and I take them up in later chapters.

In March 1991 the draft of a report from the NIH Office of Scientific Integrity (OSI) was leaked.[32] This report concluded that the data in the *Cell* paper had been fabricated. There was strong criticism of Baltimore for having continued to defend the integrity of Imanishi-Kari and her work even as more suspicious evidence to the contrary was emerging. About a month later Baltimore issued a statement that included an apology to O'Toole and distancing himself from Imanishi-Kari. He also asked his other coauthors to join in retracting the paper. *Cell*'s editors announced that they planned to print a retraction.[33]

Commenting on the affair in the *New York Times*, John Maddox, editor of *Nature*, was also critical of Baltimore, and there have been newspaper editorials and commentary calling for more vigorous investigations by scientists of allegations within their community.[34] While the draft OSI report was being circulated for comments from the principals, there was no letup in the trading of accusations and denials in the columns of *Nature*, as Imanishi-Kari, O'Toole, Eisen, and others pursued this scientific demolition derby.[35]

Baltimore resigned as president of Rockefeller University in December 1991, saying that "the *Cell* paper controversy created a climate of unhappiness among some in the university that could not be dispelled."[36] By that time the *Cell* paper had been retracted; by July 1992 the criminal investigation of Imanishi-Kari had been dropped, and expert advice obtained by her lawyers had suggested that the Secret Service analysis had important defects. A year later *Science* carried a report of a new paper by Imanishi-Kari in the *Journal of Immunology* that appeared to validate the observations reported in the infamous *Cell* paper in 1986.[37]

With the complexity of the basic scientific issues as with the torrent of statements, denials, and attacks on the accuracy of recollections, I find it impossible to reach any sort of evaluation of the merits of either side in this unhappy case. Some general comments, however, are certainly in order. As was asserted early in this case, there can be honest differences in interpretation of a given set of observations. These differences can stem from the sometimes subjective evaluation of the importance of various factors, and this is a normal part of research. Opposing views can be strenuously expressed and resolution may be impossible until new experiments are carried out, but none of this need be seen as fraud.

Massaging the Data

Blondlot's imagination produced unfortunate results, which are now regularly but uncritically included in compilations of cases of scientific fraud. The case of Robert Andrews Millikan is also often cited but is less easily classified. Millikan distinguished himself through experiments carried out first at the University of Chicago and later at the California Institute of Technology. In Chicago he devised a method for the precision measurement of the electric charge carried by the electron and also produced the first accurate confirmation of Einstein's 1905 theory for the photoelectric effect that had been discovered by Philipp Lenard in 1902. For his elegant experiments, Millikan was awarded the Nobel Prize in 1923, two years after he had moved to Cal Tech. By then his research interests were changing, and he became an important contributor to cosmic-ray explorations.

Millikan's experiments in Chicago were important components in the grand march that established modern physics, starting just before the turn of the century. In 1900 Planck had shown that the properties of heat radiation could be very accurately described if a radical assumption were granted: that the radiation being emitted and absorbed did not consist of waves (as had been thought) but rather of bundles (quanta) of energy. The familiar wave description still applied to the radiation in its transit between emission and absorption.

This departure from the old view of the smoothly continuous nature of radiation had a counterpart in measurements of electric charge. Simple measurements show that electric currents seem to behave like fluids and can be smoothly adjusted in strength. It had long been known that an electric charge could be produced by rubbing certain materials together and that an electric current consisted of the motion of electrically charged objects. But many of the properties of electric charge remained unexplained: Could the charge come in any amount; could it be changed smoothly? What Millikan showed in an experiment first reported in 1910 was that there was indeed a smallest possible charge. All electrons had identical charges, and no smaller charge existed in nature. This fundamental charge is so small that an electric current, made up of flowing electrons, has so many electrons that the lumpiness of the electric current is masked. (Your electric toaster uses a flow of around 10^{20}—one hundred million trillion—electrons each second.)

In his famous experiment (now a standard undergraduate laboratory exercise) Millikan measured the slow movement of very small electrically charged oil drops, following their movements through a microscope. Using X-rays, Millikan could change the electric charge on a drop and then see how its movement changed. The changes in a drop's electric charge always turned out to be exact multiples of what we now call the charge on the electron. Millikan's experiment has been repeated with improvements, and the size of the electron's charge has also been deduced from other experiments. The fundamental nature of the electron's charge has been incorporated into atomic physics, and the theory of the chemical bonding of atoms to form molecules. There is ample reason to consider this as being as close to a scientific "fact" as we have.

However, in recent years the theory of nuclei and subnuclear particles has been built on smaller units of one-third and two-thirds of the electron charge. Quarks, carriers of the fractional charges, are considered to be fundamental building blocks of other particles. Their existence has been deduced from many experiments, but "free quarks," outside normal nuclei, have not been seen. There is as yet no evidence anywhere either for fractional charges other than one-third and two-thirds or for a continuous range of charges.

Researchers have wondered whether Millikan might have seen any sign of free quarks. Some of his notebooks still exist, though none from the important year of 1909. Gerald Holton and Allan Franklin have examined the notebooks for evidence of fractional charges, which would have shown up in the anomalous movements of some drops. None has been conclusively identified, but something else turned up: Millikan was selective in the data he reported. Measurements were annotated as "best" and "very good"; some droplets yielded ratings of "fair."[38] Millikan discarded data from some drops because they "yielded values ... from two to four percent too low."[39] Other drops were applauded: "Exactly right. ... Publish this one. Beautiful. BEAUTY one of the very best."[40] Data from one drop were excluded because they yielded a value about 10 percent too low, and Millikan decided that the object he had observed "could not have been an oil drop," perhaps a dust particle instead.[41]

Holton suggested that Millikan had accepted the idea of the specific nature of the electron charge and that this was a "mechanism necessary for stabilizing belief in efficacy of [the] hypothesis long enough to help it to survive to the later stage of testing in public discussion."[42] In other words Millikan was working to test a hypothesis (that all electrons had identical charge) and believed in this so strongly that he set aside discrepant results.

At the same time Felix Ehrenhaft in Vienna was carrying out similar experiments and finding all sorts of values for the electron charge. Holton described the competitive nature of the research by Millikan and Ehrenhaft and suggested that this colored their approaches. But what is one to make of Millikan's subjective acceptance and rejection of data? William Broad and Nicholas Wade have included Millikan among the case studies in their book *Betrayers of the Truth*, with the strong implication that Millikan was dishonest yet won the Nobel Prize, while his honest competitor, Ehrenhaft, received no recognition despite several debates on the conflicting results.

This sort of indictment is misleading, perhaps unfair. Millikan did indeed establish very precisely the value for the electron charge, but there had been earlier experiments that had indicated quite strongly that the electrons should all have the same charge, though the precision was not high. Millikan and Ehrenhaft were really testing this hypothesis, not starting from absolute zero. Any working scientist knows that there is likely to be a background of measurements that can emerge for all sorts of instrumental reasons and need to be discarded.

It is notable that some scientists do have or develop a sixth sense, a gut feeling for what is correct or likely to be fruitful and for what to mistrust. This feeling can, and has, led people astray, and Millikan was later conspicuously wrong in his theories of the type and origin of cosmic rays. But with the electron's charge he was correct. Is it fraud to get the correct answer at a time that the correct value was not known with great precision? Fudging to get agreement with a known value is something else, but that was not the case here. Science was not misled or held back by Millikan's analysis. The fundamental particles of nature (quarks excepted) *do* have no less than the electronic charge. It would probably have been better if Millikan had reported all of his data, then selected those used for his best average value, with commentary as needed. Perhaps he was lucky. Presenting only the good data runs the risk that the definition of *good* may be wrong, incomplete, or change with time, but I do not feel that Millikan's behavior is truly comparable to that of researchers who have fabricated data from start to finish.

A passing comment: Old-fashioned notebooks permit this sort of archival detective work, but this may now be difficult or virtually impossible with computer-stored data. Electronic signals are recorded and manipulated long before hard copy appears; after even a few years, it can turn out that there is no one around who knows just what computer program was used. Some of my colleagues are throwing away thousands of IBM punched cards, for we no longer have the hardware to read them. Experimental setups are so complicated that only the scientists who built the apparatus and analyzed the data will know of their idiosyncrasies. In future years our checks may lie not in looking at notebooks or trying to decipher computer tapes but may require repeated experiments, and that might well be impossible because of the costs or because of the variable nature of the phenomena being observed.

A Verdict

Science is certainly far more complex than popular impression would have it. Some fraud has definitely occurred, but these incidents represent a very tiny fraction of the total scientific endeavor. As Stephen J. Gould has argued, it is important to separate honest error from fraud.[43] In today's climate, he has suggested, the Medici might have conducted hearings to investigate Galileo for fraud, for his erroneous claim that Saturn had two moons. There are cases, such as those of Burt and Millikan, where the verdict is still open. Should Blondlot be accused of fraud? I do not think so. What about Pons and Fleischmann? Fraud—definitely not; sloppy science—definitely yes. We return to these questions after one more chapter on unconventional science.

Notes

1. L. Hearnshaw, *Cyril Burt, Psychologist* (Ithaca, N.Y.: Cornell University Press, 1979).
2. Ibid.
3. *Science* 201 (1978): 1177.
4. *The Burt Affair* (New York: Routledge, 1989); see review in *Isis* 83 (1992): 698.
5. *Science* 340 (1989): 439.
6. *Science, Ideology and the Media: The Cyril Burt Scandal* (New Brunswick, N.J.: Transaction Publishers, 1991).
7. *Times,* November 10, 1976, 17.
8. *Cyril Burt,* 243.
9. *The IQ Controversy* (New Brunswick, N.J.: Transaction Publishers, 1988), 89.
10. Ibid.
11. Thomas J. Bouchard Jr. et al., *Nature* 250 (1990): 223.
12. *Nature* 70 (1904): 530.
13. William Seabrook, *Doctor Wood* (New York: Harcourt, Brace, 1941), 238.
14. "N-rays," in *Historical Studies in the Physical Sciences* 11 (1980): 125.
15. *Minerva* 27 (1989): 177.
16. *Nature* 325 (1987): 207.
17. William Broad and Nicholas Wade, *Betrayers of the Truth* (New York: Simon and Schuster, 1982).
18. *Nature* 286 (1980): 437.
19. Ibid., 294 (1981): 509.
20. *Science* 134 (1961): 945–946.
21. Ibid., 234 (1986): 1056; see also 1069.
22. Ibid., 211 (1981): 1022.
23. *Nature* 338 (1989): 613.
24. Ibid., 341 (1989): 11, 13.
25. Ibid., 16.
26. Ibid., 343 (1990): 305.
27. *Journal of the Geological Society,* January 1991.
28. *Nature* 355 (1992): 660.
29. *Cell* 45 (1986): 247.
30. *Science* 244 (1989): 643.
31. *Issues in Science and Technology* 5 (summer 1989): 48.
32. *Nature* 350 (1991): 259, 262, 269, 563.
33. Ibid., 351 (1991): 85, 94; *Science* 252 (1991): 768.
34. *New York Times,* March 31, 1991.

35. See, for example, *Nature* 351 (1991): 180, 341, 691; 352 (1991): 101, 183, 514; and *Science* 253 (1991): 24.

36. *Science* 254 (1991): 1447.

37. Ibid., 260 (1993): 1073.

38. *Historical Studies in the Physical Sciences* 9 (1978): 193.

39. Ibid., 202.

40. Ibid., 212.

41. Ibid., 11 (1981): 193.

42. Ibid., 9 (1978): 213.

43. *New York Times*, July 30, 1989, and *Applied Optics* 28 (1989): 3851.

11

Political Pseudoscience

Over the centuries many scientists have chosen to become active or found themselves caught up in politics and sometimes have paid a high price. Galileo was tried by the Inquisition in 1633 and found guilty of disregarding an earlier instruction to desist from teaching the Copernican thesis that the earth moved rather than stood still. This case was notable for the intrusion of a temporal power into the content of science. There have, of course, been numerous instances of scientists' being penalized for their political views or activities. When Antoine-Laurent de Lavoisier, the great French chemist, was guillotined in 1794, the president of the tribunal that sentenced him was reported as commenting, "The Republic has no need of scientists; justice must take its course." But there was no attempt to argue with his scientific discoveries. Lest we think that such events are far behind us, let us recall the McCarthy years and the treatment of J. Robert Oppenheimer and, more recently, the fate of Fang Li Zhi, the exiled Chinese astrophysicist. What distinguishes these political cases from those I describe in this chapter is the reversion to the Galileo case, with governments attempting to choose between competing scientific theories to dictate what constitutes correct or acceptable science.

In most of the cases we have examined in previous chapters, the innovators have generally had scientific motivations or have made claims to be scientific even if there were lapses from the required standards. But sometimes the running is made by forces external to science, and the results have been uniformly bad and sometimes disastrous. We saw some of this in the AD-X2 battery additive case. That transparent attempt to assert the correctness of a claim was benign in comparison to the cases that I now describe, where political or theological agendas provided the motivating forces. No new science has resulted, but the scientific damage has sometimes been long lasting. The lesson is that science must continually fight to prevent itself from being used for nonscientific ends. Of course, society will decide what to support in scientific research and how it will use the results of science. Scientists as informed citizens must take part in these debates, but none of this should distort the content of science, the actual findings or interpretation of scientific results. Those findings must emerge from within science and not be dictated from the outside. The examples that follow will, I think, make this argument with force.

The Lysenko Affair

The similarities between children and their parents do not really surprise us. With the advent of photography, family resemblances can be traced over many generations where previously we had to rely on descriptions. In addition to appearances, traits such as extreme mathematical or musical ability run in some families. Though in some traits the genetic influence, as we now describe it, is undeniable, the influence of the environment has also long been seen. Some political policies have hinged on tilts in this nature-nurture debate, as we have seen in the strange case of Cyril Burt.

We find interesting parallels when we move from the discussion of traits of humans to those of plants. Many factors have contributed to the growth of the world's population and to a general increase in longevity; one of these is the domestication and improvement of plants used for food. The modern science of molecular genetics now permits the development of fruits and vegetables with all sorts of desired characteristics, as well as the promotion of some varieties whose impressive appearances substitute for flavor.

Long before the molecular basis of genetics was established, agronomists had discovered how to improve many plant species through selective fertilization and propagation. After the Russian Revolution, there was an urgent need to increase agricultural productivity, and it was recognized that improved wheat strains were required. There were two competing research programs to achieve the desired results. They were headed by Nikolai Vavilov and Trofim Lysenko. Vavilov was born in 1887 into a family wealthy enough to send him to England for his education. He visited research laboratories in England, France, and Germany and returned to Russia at the start of World War I. Though not a Communist, he supported the aims of the revolution and later the Soviet government. By 1922 Vavilov was placed in charge of agricultural research, becoming director of the Institute of Applied Botany. Based on Mendelian theories of genetics, his research involved the slow process of following generations of plants, selecting successively those that displayed desired characteristics. His work was widely respected by the Lenin government and also initially by Stalin, as well as by scientists both in the Soviet Union and outside. Vavilov was an inspired speaker and leader. By 1929 the Lenin Academy of Agricultural Sciences, under his leadership, was a powerful confederation of research centers. Genetics research in the Soviet Union was highly regarded and attracted scientists from the West, among them H. J. Muller, who would later return to the United States, where his work won him a Nobel Prize.

While all of this was going on, there was a competing approach to plant improvement that truly qualifies as pseudoscience. This method, following the ideas of the early-nineteenth-century scientist Jean Baptiste de Lamarck, was based on the plausible assumption that characteristics acquired during the lifetime of a plant (or animal) could be transmitted to the next generation. So, for example, it was thought that a giraffe that stretched its neck to reach the topmost leaves would develop a trait that would continue in its progeny. This theory, though discarded in the West, was taken up in Russia by I. V. Michurin (1855–1935). (There is a town southeast of Moscow still named Michurinsk.) After the revolution "progressive Michurinist biology" was adopted by Lysenko, who made extravagant promises and boasted of the speed with which he could develop improved strains of wheat. Lysenko was born in the Ukraine in 1908. His family were peasants; he was educated as a practical agronomist and does appear to have had some talent (as did Luther Burbank in the United States, whose ideas also fell well outside the range of those generally accepted as scientific). Lysenko's early research related to sugar beet and tomato breeding, but in the mid-1920s he turned to a study of the temperature dependence of plant growth and from this developed his method of "vernalization." When he first used this procedure, he required that winter wheat seeds be moistened and chilled then planted in the spring. In this way winterkill would be diminished if not eliminated. His father tested this system as early as 1929, and the results were claimed to be so successful that the Ukrainian Commissariat of Agriculture expanded the experiment to 1,000 test plots for 1930. The technique was gradually streamlined to the simple planting of perfunctorily moistened seeds. Laboratory tests were undertaken but were similarly shoddy, and in one trial Lysenko claimed success for converting winter wheat to spring wheat on the basis of the observed germination of a single plant, with no controls.

The subsequent contest between the programs of Vavilov and Lysenko has been chronicled in great detail in two books. In 1969 *The Rise and Fall of T. D. Lysenko* by the Russian scientist Zhores Medvedev was published in English in the United States after Medvedev could not obtain permission to publish the book in the Soviet Union.[1] Medvedev was at that time head of the Laboratory of Molecular Biology at the Institute of Medical Biology in Obninsk. In 1973 he left the Soviet Union; his passport was then revoked, and he settled near London, working in the laboratories of the National Institute for Medical Research. David Joravsky's book, *The Lysenko Affair*, came out in 1970.[2] Together, these books document the remarkable history of the way in which the Soviets' desperate quest for expanded agricultural production led to the championing and elevation of the charlatan Lysenko and the elimination of a great scientist, Vavilov. The scientific merits of the two sides had been known in the West for many years. As early as 1951 Lysenko's contributions were succinctly described by Theodosius Dobzhansky of Columbia University:

Neither Lysenko nor any one of his numerous followers have produced a single new or original idea, either a right or a wrong one. It can be stated without hesitation that michurinist biology is nothing more than a relapse towards views that were current in biology in the nineteenth century, and which were discarded early in the present century, mainly owing to the discoveries of genetics. The sum total of what the lysenkoists have to offer is abandonment of the chief attainments of biological research and thought during the current century.[3]

Even before the appearance of Medvedev's book, the general outlines of the sequence of events were also known, but the details were both revealing and horrifying.

In 1931 the government published a decree that posed problems to the Lenin All-Union Academy of Agricultural Sciences (LAAAS) and the All-Union Institute of Plant Breeding (AIPB). The problems dealt with developing varieties of cereal crops adapted for different climatic regions; the decree demanded that the ten years normally required for such development be shortened to four years. Vavilov, then president of the LAAAS and the AIPB, understood that the demands could not be met, but Lysenko published a pledge to produce the needed varieties in two and a half years.

There followed increasing favors and honors for Lysenko. At a 1935 congress, Stalin applauded Lysenko. As Medvedev wrote, "And after Stalin's 'Bravo, Lysenko, Bravo,' a new and special period began in Lysenko's activities and in the history of Soviet biology."[4] Vavilov was arrested in 1940; in the following year he was found guilty of belonging to a "rightist conspiracy" and of spying for Britain and sentenced to death. He died in prison in 1943. In contrast, Lysenko became director of the Institute of Genetics. His influence extended even into high school biology textbooks, from which descriptions of Mendelian ideas were eliminated.

Stalin's support of Lysenko went beyond simply backing him and eliminating the competition as identified with Vavilov. Lysenko's denial of the influence of genetics and his "demonstration" of the importance of the environment went along with the Soviet political philosophy. But as far back as 1906, Stalin had discussed Darwin's theory in an article, and he now took an active role in using Lysenko as a vehicle for his ideas on Darwin and society. With the collapse of the Communist state, scholars have been able to obtain access to Stalin's papers, housed in the Central Party Archive in Moscow. Kirill O. Rossianov has found the manuscript for the speech Lysenko delivered at the 1948 meeting of the LAAAS.[5] The manuscript contains Stalin's handwritten contributions—comments, corrections, deletions, and additions. The tenor of Stalin's editing was to play down Lysenko's emphasis on class-based science, and he deleted all uses of *bourgeois* as a qualifier of *science*. In his unsigned annotations he also expressed his support for Lamarckism.

Thus, at that 1948 meeting of the Lenin Academy, Lysenko was still being hailed for his scientific work, and "classical genetics was denounced as contrary to Darwinism, Michurinism and dialectical materialism."[6] Nevertheless, criticism of Lysenko's ideas and claimed results increased, though he was able to continue through protection by friends in high positions. Nikita Khrushchev visited Lysenko's experimental farm in the Lenin Hills (near Moscow) and praised Lysenko's work. But already in 1955 a widely circulated petition called for his removal from the presidency of the LAAAS. Non-Lysenko genetics continued to recover, though for some years only Lysenkoists were given permission to attend conferences outside the Soviet Union. During those years Lysenko supported Khrushchev's agricultural policies and in return was able to continue his work on the Lenin Hills farm, where he crossbred cattle to try to improve the butterfat content of their milk. An official study of his results, however, showed that far from bringing about any improvement, the bulls that had been produced and widely used for further breeding had ruined many herds where they were used.

In 1964 the elections to the Soviet Academy of Sciences provided a base for further critical review. Lysenko was removed from directorship of the Institute of Genetics. By 1974 the extent of the changes could be seen in a decree from the Central Committee of the Communist party, which discussed "measures to accelerate the development of molecular biology and molecular genetics and the use of their achievements in the national economy" and charged the Academy of Sciences with designing appropriate research programs.[7] Even so, the thaw continued to be slow. In 1979 a biography of Vavilov appeared in the Soviet Union, its English translation in 1987.[8] In it I find no mention of Lysenko. Vavilov's early career is described in fulsome terms, but there is not the slightest hint of the troubles, still less of Vavilov's fate. The book ends abruptly, a few pages after reporting a 1940 address to a Young Pioneers' rally, described as Vavilov's "last public speech."[9] In a recent collection of essays, Mark Adams set the Lysenko affair in the broader context of the nature-nurture debate within the Soviet Union.[10] The parallels and contrasts with corresponding debate in the West make interesting reading but drift far from my main theme of differentiating between science and its imitators.

As previously closed archives become open in the Soviet Union, the Lysenko affair continues to attract attention. A biography of Vavilov written in the West appeared in 1984.[11] Further material appeared in an article in *Nature*.[12] More will surely follow, but we should note a surprising inversion of what we might have expected politically. Adams, whose essay on Lysenko I cited above, noted the strong political undercurrent in the debates over the hereditary versus the environmental influence on measures of intelligence. In general in the West, the role of the environment has been stressed by people on the political left, while the supporters of genetic dominance have tended to be much more conservative politically. The inverse has been the case in the Soviet Union, as shown by the Lysenko affair.

Aryan Physics

At about the same time that Lysenko's pseudoscience was being given official support in the Soviet Union, another political incursion into science was taking place in Germany. During the Nazi years "Aryan physics" was championed; attempts were made to suppress relativity and quantum theory, labeled as "Jewish physics." When one considers how methodically the Nazi machine usually worked and how meticulously efficient it was in its documentation, its assault on the core of modern physics was surprisingly clumsy and halfhearted.

Science, especially in the United States, was enriched in the 1930s by the stream of Jewish refugees forced out of Germany (and to a lesser degree out of Italy). The non-Jewish scientists who remained behind were not immune to harassment, though it was unpredictable and did not go beyond some ideologically inspired decrees that were not firmly enforced. But while such intimidation is usually associated with the Nazi regime, a pattern had been set in Germany many years before.

Germany was a broken country as it emerged from the war in 1918. The 1920s were difficult, and it is not surprising that the Germans sought scapegoats for at least some of the country's troubles. Anti-Semitism, a traditional and convenient weapon, was used to great effect. Two of the leaders of the Aryan physics movement, aimed not just to promote pride in German achievements but to do this at the expense of Jewish scientists, were Philipp Lenard and Johannes Stark. Both were experimental physicists of distinction, both Nobel Prize winners (Lenard in 1905, Stark in 1919).

The scientific targets of their anger and hatred were relativity and quantum theories—especially relativity, because of the violence that it appears to do to everyday experience. As I mentioned before (Chapter 6), there are two parts to relativity theory: The Special Theory appeared in 1905, the General Theory about ten years later. Einstein's new way of describing nature attracted attention but not comprehension. In 1908 the *Scientific American* ran a contest for the best essay to describe the fourth dimension, that strange new concept that Einstein had introduced. Even the President of the American Physical Society expressed his discomfort with this new theory.[13] When one of the predictions of the General Theory was dramatically verified by measurements made during the 1919 solar eclipse, Einstein became an instant world figure, a permanent generic icon of mathematical genius. The counterintuitive assumptions and mathematical complexity of the theory made it hard to understand, still less to accept. Even now, so many years later, relativity continues to puzzle most people and even some scientists, spurring them to denial or opposition. For example, in the mid-1950s the correspondence columns of *Nature* carried a long-running debate initiated by Herbert Dingle, at one time president of the Royal Astronomical Society. In his opposition to Einstein Dingle labored over familiar "paradoxes" but provided no new insights.[14] A seminar he presented in 1956 was a sad display of his confusion. Given the nature of the theory of relativity and its affront to common sense, this sort of rearguard resistance must be expected to continue. The German experience, though, was made very different by the involvement of anti-Semitism.

In the political climate of 1919 and despite his original support of relativity, Lenard had many reasons to dislike Einstein—his sudden fame, his pacifism during the recent war, his support for the new Weimar government, even his role as a theorist. A debate on relativity theory was arranged for the 1920 meeting of the German Physical Society at Bad Nauheim. Planck, originator of the quantum hypothesis, presided at the relativity session. Because of concern that the discussion not be diverted from scientific issues, the meeting was tightly organized, but there was some rowdiness when Einstein spoke, and the hostility directed to him increased in subsequent years.

In 1931 another denunciation appeared, *One Hundred Authors Against Einstein*, with essays from otherwise competent scientists venting their prejudiced views; one contributor, a professor in Minneapolis named Arvid Reuterdahl, labeled Einstein "the Barnum of physics."[15] After Jewish university faculty were quickly ousted from their positions when the National Socialists came to power in 1933, the campaign against relativity and quantum theory persisted.[16]

The philosophical base of the campaign against Einstein (and others) was clearly set out by Stark in an article that appeared in *Nature* in 1938. Stark tried to draw a distinction between "pragmatic" scientists, whom he praised for being "directed towards reality" as opposed to scientists of the "dogmatic school." A "dogmatic scientist," he said,

> starts out from ideas that have arisen primarily in his own brain, or from arbitrary definitions of relationships between symbols [and] by logical and mathematical operations he derives results in the form of mathematical formulae. ... The dogmatic spirit leads to the crippling of experimental research and to a literature which is as effusive as it is tedious. ... I have directed my efforts against the damaging influence of Jews in German science, because I regard them as the chief exponents and propagandists of the dogmatic spirit.[17]

The campaign against Jewish physicists had expanded to include even non-Jews who used relativity or quantum theory. So, for example, in 1937 there was a vicious attack on Heisenberg in the SS journal *Das Schwarze Korps* (The black corps). A series of three articles, probably inspired by Stark, described Heisenberg, one of the founders of quantum mechanics in the 1920s and a Nobel Prize winner, as a "white Jew" because of his use of the odious theories. Heisenberg knew full well that modern physics could not exist or progress without these theories. Following an event of somewhat surrealistic dimensions, Heinrich Himmler, chief of the SS, eventually intervened on Heisenberg's behalf. Heisenberg's grandfather and Himmler's father had known each other as principals of secondary schools in Munich, so Heisenberg's mother paid a call on Himmler's mother, who promised to take up the matter: "My Heinrich ... is such a nice boy. ... If [he] only knew of this, then he would immediately do something about it."[18] And so he did.

Although German physicists continued to use relativity and quantum theory, as Heisenberg had been instructed, they were careful not to give any credit to Einstein; in fact, for several years his theory was attributed to others. The quality of German physics deteriorated sharply after 1933. There were many contributory factors: the expulsion of Jewish scientists, the departure of others who decided not to remain in academe, and a major decrease in the number of students studying physics. In those years there were, of course, more pressing problems for the government to deal with, but the quality of German science was significantly diminished, and its recovery after the war was slow.

Creation Science

When Stalin backed Lysenko, he compounded the damage by permitting the elimination of Vavilov. Soviet agriculture and biological science paid a terrible price. The German attempts to suppress the theory of relativity formed only a minute part of a sweeping program of anti-Semitism. Beyond the loss of so many outstanding scientists, a major consequence of the purging of the German universities was the elevation of second-rank scientists, rewarded for their party loyalty. But the content of science was not seriously damaged, since the proscription of relativity and quantum theory could be evaded.

In contrast, the attempt to suppress the teaching of the theory of evolution in the United States was far more successful, though it did not bring personal attacks of the sort that had taken place in the Soviet Union and Germany. Though nationwide in its effects, the opposition to evolution in the United States was not a truly national movement but rather has flourished in those states where fundamentalist churches have the greatest influence. A significant reason for the antievolution successes is that education is largely a state and local concern, making it easier to exert control. In addition, when tests of the constitutionality of the antievolution legislation came to local courts, they encountered judges who were sensitive to the views of the voters they needed for reelection. Conversely, the antievolution forces have suffered major defeats when litigation reached the federal court system, which was better insulated from local politics. As in Germany, a nonscientific agenda fueled these attempts to dictate what should be accepted and taught as science.

At the center of the antievolution sentiment is the unquestioning belief in the literal truth of the Bible. From this there follows the equally rigid acceptance of the account of the Creation as described in the opening chapters of Genesis as well as a number of deductions that have been drawn from them. These include a relatively modest age of the earth, variously estimated at between 6,000 and 10,000 years and the absence of significant evolution, certainly as it relates to humankind. These beliefs are not new, and there have been many attempts to calculate the age of the earth by tabulating the number and duration of generations listed in Genesis. In this way Luther arrived at a date for the Creation around 4000 B.C., and in 1650 Bishop Ussher of Armagh placed it at 4 P.M. on Sunday, December 23, 4004 B.C.

The start of the scientific quest to measure the age of the earth can reasonably be set in the eighteenth century with the attempts to reconcile the observations of fossils with the growing body of geological knowledge. As more sophisticated methods came into use, the estimates of the earth's age produced steadily increasing values. Today, with sensitive measurements of radioactivity and with our knowledge of nuclear physics, we date the formation of the earth and the rest of the solar system to close to 4.5 billion years ago, with an uncertainty of no more than 1 percent.

Since the publication of Darwin's *Origin of Species* in 1869, paleontology and evolutionary biology have developed a picture in which the origins of humans can be placed a few million years ago. The evolution of many plants and animals can sometimes be traced in detail and usually in broad outline. Experts disagree vigorously on many of the particulars, but there is little scientific dissent from either the general idea of evolution or the associated time scale.

The development of this evolutionary view of the history of the earth together with its living organisms has altered the type of scientific question that is posed. Two hundred years ago the question was, Within the accepted 6,000 years, how could geological strata have been formed and fossils have been produced? Now we ask, How long must it have taken to produce the observed strata, the fossils, the relative abundances of the various isotopes of lead formed in the radioactive decay of uranium and thorium atoms? A central feature in the modern approach is the acceptance of uniformitarianism, the belief that the earth that we see has been shaped mostly by processes that we recognize today, acting over many years at the same speeds that we can measure now (see Chapter 3). We assume that the chemical reactions and physical processes did not behave differently in the past. While this is indeed an important assumption, there are some experimental checks. When astronomers look at distant stars and the much more distant galaxies, they are detecting light that has been traveling for millions and billions of years. When we use this old light to interpret events in those places so distant in space and in time, we get agreement with what we can see taking place much closer. We thus have every reason to think that the general idea of uniformitarianism is probably highly accurate and at least a good guide.

Religious belief can be very strong; it has been a powerful force in human history and will surely continue. Faced with the growth of scientific knowledge, many denominations have adjusted their views, accepting the findings of science while preserving their fundamental faith. The separate roles for science and religion were well described by Galileo: "The intention of the Holy Ghost is to teach us how one goes to heaven, not how heaven goes."[19] Not surprisingly, such acceptance of science has not been unanimous. A belief in the literal inerrancy of the Bible and its absolute validity in calculating the age of the earth and the rejection of the idea of evolution—these form a common thread among fundamentalist religions. But even beyond the membership of these organizations, there is a widespread acceptance of such views, providing a broad base for opposition to the teaching of evolutionary theory.

Within the United States this opposition took the form of state legislation that prohibited the teaching of evolution. After Tennessee passed just such a law in 1925, the American Civil Liberties Union (ACLU) offered to support a court test of the law's constitutionality and found a willing participant in John Thomas Scopes, a biology teacher who introduced evolution into his Tennessee classroom and was prosecuted. The flamboyant personalities of William Jennings Bryan, the prosecutor, and Clarence Darrow, defense counsel, guaranteed a degree of national attention far greater than the trial would probably have attracted on its own merits. Scopes was found guilty. Although Scopes's appeal succeeded on technical grounds, evolution was not really the winner, and the impact of this case has been well documented.

Surveys of the content of high school biology textbooks have shown how in the forty years following the Scopes trial, the treatment of evolution was systematically reduced or eliminated.[20] This process has been traced through successive editions of widely used texts and by inquiries addressed to science editors at some of the major publishing houses. In many states textbooks are adopted by local school districts from lists of approved books compiled by state boards of education, and state legislatures or strongly antievolution school boards ensured the virtual elimination of evolution from more than a generation of high school texts. Although some publishers had yielded to antievolution pressure before the Scopes trial, the failure of the direct challenge to the Tennessee law gave a considerable boost to the move against the teaching of evolution. It took until 1967 for Tennessee's Butler Act to be repealed, but a significant revival of opposition to the obscurantist legislation had emerged in the earlier 1960s. This was a time of broad and critical national review of the high school science curriculum, following the United States' embarrassment when the Soviet Union launched the first artificial earth satellite, *Sputnik*, in October 1957. The findings highlighted the generally sorry state of high school science. (One might wonder whether the situation is any better today, more than thirty years later.)

In 1968 a challenge to a ban on the teaching of evolution finally reached the U.S. Supreme Court. Susan Epperson, a biology teacher in Little Rock, filed suit, testing the constitutionality of the Arkansas Rotenberry Act after the Arkansas Supreme Court upheld the legislature's right to determine the high school curriculum. The U.S. Supreme Court held that the Arkansas law did effectively support a particular religious doctrine and was thus in contravention of the First and Fourteenth Amendments: "It is clear that fundamental sectarian conviction was and is the law's reason for existence."[21]

Until this time the goal of the antievolution forces had been the elimination of evolution as a curriculum topic, and in this they had largely succeeded until the Arkansas decision. We see, however, a shift in tactics from the mid-1960s, with the formation of the Creation Research Society. The new approach was to proclaim the fundamentalist view a part of science and to name it *creation science*, worthy of being considered as a scientific alternative to evolution. A component of this attempt to disguise fundamentalism was to recruit the support of scientists in order to give the appearance of scientific legitimacy, at least to those who were not sufficiently knowledgeable. For the Scopes trial, Bryan had tried to gain antievolution support from scientists but could name only one person, a self-described geologist who had neither field experience nor recognized training. The creation science movement, however, is proud to advertise the support of many scientists with Ph.D.'s. Closer inspection shows that a good number have their degrees in fields not particularly relevant to the subjects being debated, though some do hold tenured faculty positions in major universities. There is certainly enough of the appearance of scientific support for creation science to feel secure in its claim for equal time with evolution in the curriculum and textbooks. Accordingly, the creation science legislative attack has taken the form of requiring that evolution not be presented as fact but rather as "only a theory," and on this basis, if discussed at all, then only in conjunction with creation science. In an article in the *Yale Law Journal,* Wendell R. Bird described another argument that has been put forward: "Exclusive public school instruction in the general theory of evolution at the secondary and elementary levels abridges free exercise of religion."[22] Bird is on the staff of the Institute for Creation Research (ICR) in California, the intellectual center of creation science. It publishes books and a newsletter (*Acts and Facts*) and provides speakers for debates with scientists.

Prominent figures in the creation science movement have attempted to give scientific justification to their claims that the earth is relatively young. They have subjected evidence for the generally accepted solar system age of 4.5 billion years, for the age of fossils, and for evolution in general to exhaustive scrutiny and have claimed support for their ideas on the basis of minuscule debating points that leave the major structure of evolution untouched. The futility of this sort of "scientific" debate is well illustrated by the last line of creation science defense: When confronted with the vast range of evidence testifying in a consistent way to the age of the earth, the solar system, and indeed our universe, proponents of creation science respond that all of these things were created just 6,000 years ago *with the appearance* of great age. So, for example, light waves we see today did not really start out from the Andromeda Nebula 2 million years ago but were set in motion much closer to the earth when the universe was created, so as to make us *think* that they had been traveling for that great time. It has been suggested that following this line of argument, we could just as easily consider the universe to have been formed only ten minutes before you started to read this chapter, with all of us created with our memories in place, coffee already made. The possibilities are endless.

As I found with Velikovsky's claims for scientific support for his theories and predictions, so with creation science. Each of their claims needs to be examined with care, for their partisan shading of evidence is not always obvious. For example, in their book *What Is Creation Science?* Henry Morris and Gary Parker claim that Hannes Alfven, a physics Nobel Prize laureate, has "shown conclusively that ... the Big Bang is only one of several possible explanations" for certain radio astronomy findings.[23] Those radio observations, reported in 1965, are generally considered to have provided very strong support for the big bang theory of cosmology. Reference to the Alfven letter in *Nature*, however, reveals a comment that Morris and Parker carefully omitted from their quotation. Alfven and his coauthor took issue with one interpretation of the radio measurements, then noted that "we do not claim that the observed microwave background radiation is inconsistent with either the basic concept of a hot big bang cosmology, or with a completely different class of evolutionary cosmology."[24] This sort of manipulation of the evidence by creation scientists has been repeatedly documented.[25]

Even though the overwhelming majority of scientists can see the weakness of the creation science case, many state legislators were moved by other concerns, and Governor Frank White signed Arkansas Bill 590 into law in March 1981 to require equal treatment of both evolution and creationism. A suit challenging its constitutionality was soon filed by a broad coalition of church leaders (United Methodist, Episcopalian, Roman Catholic, African Methodist Episcopal Churches, Presbyterian, and even some Southern Baptist clergy). The plaintiffs included a high school biology teacher, a number of Jewish groups, and the National Association of Biology Teachers. As with the Scopes trial, this case attracted considerable publicity. Unlike the Scopes trial, this case was notable for the parade of expert witnesses. The creation science defense called to the stand seven supporters, including Henry Morris and Duane Gish, respectively director and associate director of ICR. Counsel for the opposing side called on philosophers of science, theologians, and such scientists as paleontologist Stephen Jay Gould of Harvard. Through the many days of testimony and exhaustive cross-examination, it was made clear that the case for creation science rested on a set of beliefs particular to specific religious groups and that creation science failed to meet the requirements normally expected of science. In particular, where science responds (sometimes slowly) to demonstrations of error or omission, creation scientists emerged as totally impervious to all and every criticism, their theory unfalsifiable, the very antithesis of a genuine science. It was notable that the creationists called no significant religious leaders or theologians as witnesses.

The language of Arkansas Senate Bill 590 is the same almost to the letter as that which many other states have considered. In a doublespeak that George Orwell might have admired, Bill 590 starts by setting out the purpose of the proposed law: "to require balanced treatment of Creation-Science and Evolution-Science in

Public Schools; to protect Academic Freedom by providing student choice; to ensure freedom of religious exercise; to guarantee freedom of belief and speech; to prevent establishment of religion." In Section 4 the definition of *creation science* is breathtaking in its sweep:

(a) "Creation-science" means the scientific evidences for creation and inferences from those scientific evidences. Creation-science includes the scientific evidence and related inferences that indicate: (1) sudden creation of the universe, energy, and life from nothing; (2) The insufficiency of mutation and natural selection in bringing about development of all living kinds from a single organism; (3) Changes only within fixed limits of originally created kinds of plants and animals; (4) Separate ancestry for man and apes; (5) Explanation of the Earth's geology by catastrophism, including the occurrence of a worldwide flood; and (6) A relatively recent inception of the Earth and living kinds.

Further (Section 5), "The Act does not require or permit instruction in any religious doctrine or materials ... [but] simply requires instruction in both scientific models (of evolution-science and creation-science) if public schools teach either." The "Legislative Declaration of Purpose" (Section 6) asserts that its purpose is "protecting academic freedom" and "preventing establishment of Theologically Liberal, Humanist, Nontheist, or Atheist religions." In the "Legislative Findings of Fact" (Section 7) the act states, "Evolution-science is not an unquestionable fact of science, because evolution cannot be experimentally observed, fully verified, or logically falsified, and because evolution-science is not accepted by some scientists" and "creation-science is an alternative scientific model of origins and can be presented from a strictly scientific standpoint without any religious doctrine." And so it goes. It is worth remarking that there seems to be an inconsistency: On the one hand, Bill 590 argued that creation science should be treated as science; on the other, Bird earlier urged that the exclusion of creation science "abridges the free exercise of religion."

The inspiration for this form of legal approach clearly comes from the creation science movement. As far back as 1973, a very similar bill was introduced into the Georgia legislature but never adopted. The Georgia bill (H.B. 859) gave a sweeping definition to the theory of evolution: "The origin and development of the elements, the galaxy, the solar system, of life, of all the species of plants and animals, the origin of man, and the origin of all things are included in this cosmology." My astronomy courses would have been destroyed if I had been required to provide equal time for the creation science pseudoscience.

In January 1982 William R. Overton, a U.S. district judge, ruled the Arkansas law unconstitutional. In a model of methodical demolition of the creation-science defense, Overton found that "the Act was passed with the specific purpose by the General Assembly [of Arkansas] of advancing religion." That "creation-science is inspired by the Book of Genesis ... and ... is consistent with a literal interpretation of Genesis leave no doubt that a major effect of the Act is the advancement of particular religious beliefs." Commenting on the act, Overton referred to the "fallacious pedagogy of the two-model approach ... creation-science as defined in that Section [of the act] is simply not science." In response to the creation science claim that public school curricula "should reflect the subjects the public wants taught," the judge noted that "the application and content of the First Amendment principles are not determined by public opinion polls. ... No group may use the organs of government, of which the public schools are the most conspicuous and influential, to foist its religious beliefs on others."[26] Overton's opinion was masterful, a casebook for the analysis of most of what is wrong with creation science and the legislated attempts to use it. The plaintiffs had achieved their objective of showing that the tactics behind the adoption of Bill 590 had as their main purpose the advocacy of particular religious views.

There have been a few other cases since then, including a Louisiana case (*Edwards v. Aguillard*, 1986) that reached the U.S. Supreme Court in 1987. The Louisiana law also mandated "balanced treatment" in the presentation of evolution in the public schools. The Court, in a seven-to-two decision, found that the statute was deliberately designed to provide "the symbolic and financial support of government to achieve a religious purpose." The dissenting justices were Chief Justice William Rehnquist and Associate Justice Antonin Scalia.

No other case has yielded a verdict so clearly articulated as that of Overton. There are still local skirmishes being fought at the level of state boards of education over the adoption of texts. In November 1990 the board of education in Texas decided that high school biology texts should include evolution but not creation science. About a year earlier the State of California declined to approve an application from the ICR graduate school to be allowed to continue awarding master's degrees in science. Because procedural problems and a legal challenge by ICR delayed the implementation of the decision, these "scientific" diplomas can still be obtained. In 1992 the state reversed its position, entering into an out-of-court settlement that allows the ICR to continue under its present license, requiring only that it also present evolutionary theory. The licensing role for unaccredited schools was taken away from the California Department of Education and given to the new Council for Private Post-secondary and Vocational Education.

Some final comments are still needed on the Arkansas case and Judge Overton's decision. First, whereas the Scopes trial was heard by a jury, the 1981 case was presented before the judge alone. Given the widespread sympathy for a simple belief in the inerrancy of the Bible, it is a reasonable speculation that the decision might well have gone the other way had the trial been before a jury, especially since so much testimony involved scientific and philosophical issues. Second, the great weight of expert testimony on the nature of science, presented by the plaintiffs, was clearly a determining factor in shaping Overton's view of science, including the importance of the idea of falsifiability. As I have noted repeatedly, while demarcation may seem clear to scientists, their definition is far from being accepted by many sociologists and philosophers. A vigorous presentation from a relativist might have given the judge pause before he arrived at his decision. In particular, the plaintiffs did not complement their emphasis on the claimed lack of falsifiability (for creation science) with any demonstration that creation science had indeed been repeatedly falsified in specific detail. The phoenixlike ability to rise from the ashes of some just-demolished claim and brush itself off made creation science a moving target, but this methodological issue was never joined. Had there been a presentation of the realist-versus-relativist argument, and had the judge decided in favor of the conventional (realist) view, would this have done significant damage to the relativist industry? It seems most unlikely. But had the Arkansas and Louisiana laws been upheld in the courts, the effects on science teaching could well have been as long-lasting as the outcome of the Scopes trial.

In August 1993 the school board in Vista, California, voted to allow creation science to be taught as an alternative to evolution. The board allowed creationism to be discussed in social studies and literature classes, so that a confrontation over instruction in science classes was perhaps avoided.

For the moment, it might seem as though science has at last prevailed, though precariously, and the latest assault turned back. We should not become complacent, for the underlying problem remains. The members of the fundamentalist churches have a deep and sincere attachment to their faith; many are more than simply uncomfortable when their children are exposed in the classroom to ideas that run contrary to their most cherished beliefs. We can sympathize with them in their discomfort at the same time that we resist their remedies.

Conclusion

The cases examined in this chapter show what can happen when the content of science is dictated by nonscientific forces. The Soviet and German examples might be thought of as grotesque pathologies, products of authoritative and repressive regimes and thus not useful as models for other circumstances. The creation science issue, however, shows how science can be damaged insidiously in a more open society. To understand the reasons for the successes of these antiscientific campaigns, we need to see them against the backgrounds of their different societies. Though Michurinist biology and Aryan science are now mainly of historical interest, creation science remains an active issue.[27]

Notes

1. *The Rise and Fall of T. D. Lysenko* (New York: Columbia University Press, 1969).
2. *The Lysenko Affair* (Chicago: University of Chicago Press, 1970).
3. *Bulletin of the Atomic Scientists* 8 (1952): 40.
4. *The Rise and Fall*, 20.
5. *Isis* 84 (1993): 728.

6. E. W. Caspari and R. E. Marshak, *Science* 149 (1965): 275.

7. *Nature* 250 (1974): 7.

8. G. Golubev, *Nikolai Vavilov*, trans. Vadim Sternik (Moscow: MIR Publishers, 1987).

9. Ibid., 140.

10. Essay in Loren R. Graham, ed., *Science and the Soviet Social Order* (Cambridge: Harvard University Press, 1990).

11. Mark Popovsky, *The Vavilov Affair* (Hamden, Conn.: Archon, 1984); see reviews in *Science* 227 (1985): 1329; and *Nature* 316 (1985): 769.

12. Valery N. Soyfer, *Nature* 389 (1989): 415; see also *Nature* 335 (1990): 107.

13. In L. Pearce Williams, ed., *Relativity Theory: Its Origins and Impact on Modern Thought* (New York: John Wiley and Sons, 1968), 117.

14. *Nature* 177 (1956): 782; 178 (1956): 680.

15. Mentioned in Joseph Haberer, *Politics and the Community of Science* (New York: Van Nostrand Reinhold, 1969), 108.

16. Alan D. Beyerschen has described the machinations in German science during those dreadful years in *Scientists Under Hitler* (New Haven: Yale University Press, 1977).

17. *Nature* 141 (1938): 770.

18. *Scientists Under Hitler*, 159.

19. Galileo Galilei, "Letter to the Grand Duchess Christina" (1615), in *Discoveries and Opinions of Galileo*, trans. Stillman Drake (Garden City, N.Y.: Doubleday, 1957), 186.

20. Dorothy Nelkin, *Science Textbook Controversies* (Cambridge: MIT Press, 1977); see also *Scientific American* 234 (April 1974): 33.

21. See the article by L. Sprague de Camp, *Scientific American* 220 (February 1969): 15.

22. *Yale Law Journal* 87, 3 (1978): 515.

23. *What Is Creation Science?* (San Diego: Creation-Life Publishers, 1982), 223.

24. H. Alfven and A. Mendis, *Nature* 266 (1977): 698.

25. See R. L. Ecker, *Dictionary of Creation Science* (Buffalo, N.Y.: Prometheus Books, 1990).

26. *Science* 215 (1982): 934; see also *Academe* 68 (1982): 27.

27. The comprehensive and scholarly history of creationism by Ronald L. Numbers is essential reading for anyone concerned about this worrisome issue. See *The Creationists* (Berkeley: University of California Press, 1992).

12

Taking Stock

This book has as its central theme the separation of science from pseudoscience and the identification of correct science against a background of errors and impostors that is often present. Making these identifications and separations requires some agreement on defining what we mean by science and its methods, on what represents a scientific "fact." Within the scientific community there is a broad consensus on both of these matters, though the basis of this agreement is largely unarticulated and probably not even explicitly recognized. When we move beyond the confines of the scientific community, secure in its happy state of business-as-usual, we find a region of intense study and criticism, largely the province of philosophers and sociologists of science. Many of their writings sound strange to scientists; their concepts, language, and style of analysis unfamiliar and often apparently contorted, sometimes even hostile to science. There can be no doubt that the description and analysis of the social setting and interactions of science require the attention of these external scholars, and the size and growth of their literature attests to this. But without denying the existence of the broader context, I am concerned here with the acceptance or rejection of scientific claims, and these adjudications can be made only by scientists.

No matter how oblivious working scientists may be to the latest trends in philosophical analysis and no matter how poorly we define or understand the concepts of *paradigm* and *falsifiability*, only the scientific community will reject or accept into the mainstream of science the cold fusions, the worlds in collision, the drifting of continents. Accordingly, I do not review the externalist views to any extent; such an analysis would need to be long and would drift too far from my central concern.

While scientists, their commentators, and critics may have their various concepts of science, the much larger number of unengaged nonscientists will rarely have given the issue of demarcation any thought. The result, as we have seen, is a great tolerance for pseudoscience by the nonscientific public, often accompanied by surprise when scientists react with skepticism to the latest claims—for cold fusion, cancer cures, or whatever.

The process by which controversies of evaluation and demarcation are settled has been termed *closure*,[1] and here again we find a diversity of definitions. How does a scientific controversy end? There is no single mode of resolution. Within mainstream science, closure may be the result of persuasion, based on the clarity of the evidence, the successful rebuttal of initially plausible counterexamples or counterarguments, and the success of a theory that incorporates the new results or ideas. Such was the fate of continental drift, relativity, and quantum theory. This form of closure almost never works for pseudoscience. The devout supporters of Velikovsky, the believers in ESP and creation science, those trusting to Krebiozen cannot be swayed by rational argument or by demonstrations that adhere to the accepted norms of science. Very few supporters may be willing to change sides. Some controversial but not pseudoscientific ideas gradually fade away when confronted with repeated disconfirming evidence at the same time that clear support fails to materialize and the tide of scientific interest recedes. Examples of this category are cold fusion, polywater, and Benveniste's water-with-a-memory, but we should note that all of these involved professional scientists working within the general arena of organized science. The case of the fifth force is not truly in this latter group, for though interest has waned considerably, there are still legitimate reasons physicists would place yet lower limits on the possible strength of any new force. Creation science is unique in having a politically dispersed base from which it is still able to achieve a limited measure of local success, even while it has no scientific respectability.

To make progress, science needs facts that can be synthesized into theories. But as should be apparent by now, things are not always quite so straightforward. "Give me the facts, ma'am, just the facts," sounds like a simple instruction, but just what *are* facts? In particular, just what are scientific facts? It is easier to describe what scientific facts are *not:* They do not represent the result of a referendum among scientists, nor are facts established by the opinions of those who are not themselves scientists in the appropriate fields. Facts are not established nor are new theories accepted merely by assertion, which is usually the approach of the pseudoscientists. The acceptance of facts and theories comes by internalist consensus.

As we try to decide whether some new claim (for fact or theory) should be accepted, the style of announcement as much as the content shapes our initial responses. Professional journals carry weight; press conferences and trade books do not.

The extreme cases of eccentric pseudoscience are easily identifiable to most scientists. The ideas are often couched in newly invented terminology, sometimes with new mathematical symbols. Often there are no numbers at all. It is rare that the new proposal is supported by a test that has been carried out. Surrounding the extreme cases is a penumbra in which it is often difficult to classify research reports or innovative claims as genuinely scientific. At issue is the central question, Will some particular research, in its methods or results, survive to become an accepted part of science? The apparent sharpness of the scientific criteria is somewhat eroded by the occasional instances of dishonesty and, less frequently and in a different way, by the imposition of legislated decisions. In the end we should accept that there is a smooth shading from the genuine to the crank, with dead-end branches that distract but do not usually contribute to the advancement of science.

The subtlety of the shading is often lost on both scientists and nonscientists. I have the impression that many scientists consider it a simple matter to identify pseudoscience. This may often but not always be true. Nonscientists often find sufficient plausibility in some pseudoscientific claim that their suspicions of the motives of the scientific elite come to the fore when the scientists remain indifferent or are hostile. The problem of separation of science from pseudoscience can be addressed in different ways. In this chapter I review some tools that can help in evaluating the new reports and claims. I also describe what some scientists have done to try to clarify an often confused situation. In the next chapter I take up the question of why pseudoscience has such an enduring popular appeal.

The confusion between science and pseudoscience, between honest scientific error and genuine scientific discovery, is not new, and it is a permanent feature of the scientific landscape. Each generation will have to confront this. Acceptance of new science can come slowly. It is possible to be ahead of one's time—which is not the same as being wrong. The long delay between the publication of Wegener's continental drift hypothesis and the general acceptance of plate tectonics shows how difficult it can sometimes be to identify a scientific advance in its early stages. There are more recent examples of this sort. After about twenty years of rejection, Peter Mitchell won the Nobel Prize for chemistry in 1978; similarly, after close to forty years of being ignored, Barbara McClintock was awarded the Nobel Prize for physiology and medicine in 1983.[2]

In some ways what happened to Michael Polanyi is more interesting.[3] In 1914 and 1916 Polanyi published papers on the adsorption of gas molecules on solid surfaces. His theory received many favorable reviews, including a full discussion in a major 1922 monograph. However, further experimental discoveries and theoretical explanations led to a better theory as new results became widely known after the end of World War I. Polanyi's theory was abandoned. By 1930 further evidence had accumulated that seemed to Polanyi to rekindle support for his theory, but by then, as he put it, "the opinion that my theory was false had hardened," and he found that his work could still not gain a serious reading.[4] By around 1947 some of the major objections to Polanyi's theory had been found to be themselves in error, and his ideas were once again looked on with favor. "Could this miscarriage of the scientific method have been avoided?" Polanyi asked, answering firmly, "I do not think so. ... There must be at all times a predominant, accepted scientific view of the nature of things."[5] As has often been argued, without a paradigm scientists will be endlessly diverted; it is through the accumulation of anomalies that any errors will (even though slowly) be corrected and paradigms changed. This sort of delay has been used repeatedly by pseudoscientists and their supporters to argue against some current rejection of a crank idea. If Wegener could be right though initially rejected, then (by implication) *every* new idea labeled as wrong must also be correct—so the pseudoscientists' logic goes.

There is no definition of *science* that meets everybody's needs or avoids criticism. Similarly, demarcation of the boundary between science and its imitators has been repeatedly attempted, never to everyone's satisfaction. One well-known reconnaissance was carried out by Irving Langmuir, 1932 Nobel Prize winner in chemistry, in a talk he gave in 1953 at the General Electric Company's Knolls Research Laboratory. Langmuir's title for his seminar was "Pathological Science: The Science of Things That Aren't So." Copies of the transcript of that talk circulated for some years in samizdat fashion before appearing in accessible form in *Physics Today.* Langmuir included N-rays, ESP, and flying saucers among his targets. He was very clear that there was "no dishonesty involved" in these episodes, but people could be "tricked into false results" by many factors, such as "wishful thinking."[6]

Langmuir then set out "characteristic rules" for identifying "symptoms of pathological science":

1. The maximum effect that is observed is produced by a causative agent of barely detectable intensity, and the magnitude of the effect is substantially independent of the intensity of the cause.

2. The effect is of a magnitude that remains close to the limit of detectability; or, many measurements are necessary because of the very low statistical significance of the results.

3. Claims of great accuracy.

4. Fantastic theories contrary to experience.

5. Criticisms are met by *ad hoc* excuses thought up on the spur of the moment.

6. Ratio of supporters to critics rises up to somewhere near 50% and then falls gradually to oblivion.[7]

Another useful checklist for the identification of pseudoscience has come from Mario Bunge, a theoretical physicist and philosopher of science.[8] I paraphrase some items from his list:

1. The new theory is rigid, generally resistant to new research results.

2. Adherents generally consist of believers who do not carry out research.

3. In some cases support comes from commercial interests.

4. Most phenomena of pseudoscience are unverifiable, except by adherents, and many imply supernatural effects.

5. Supporting arguments are often drawn from outdated or discredited sources or are unverifiable. They lack clarity and consistency.

6. Use of mathematics is rare, and logical argument is often absent.

7. Many of the phenomena being claimed are very old, but there has been little or no development in ideas. (In contrast, mainstream science is cumulative.)

Bunge has drawn attention to other characteristic features: unwillingness to entertain new hypotheses and suppression or distortion of unfavorable data, even when these far outnumber the "new" results.

That veteran antipseudoscientist Martin Gardner lists "five ways in which the sincere pseudoscientist's paranoid tendencies are exhibited":

1. He considers himself a genius.

2. He regards his colleagues, without exception, as ignorant blockheads.

3. He believes himself unjustly persecuted and discriminated against.

4. He has strong compulsions to focus his attacks on the greatest scientists and the best-established theories.

5. He often has a tendency to write in a complex jargon, in many cases making use of terms and phrases he himself has coined.[9]

There is some overlap among these tests, and no particular pseudoscientific candidate will necessarily match every point on even one list. An important omission from these lists is Karl Popper's test of falsifiability. An essential component of a scientific theory is its ability to be tested. If an important deduction from the theory can be shown to be untrue, then the theory has been falsified, at least in its current form. Because of its lack of falsifiability, Judge Overton disqualified creation science in its claim to be considered a genuine science. This criterion is not always applicable, but when it can be used, it can have great force, provided that there is general agreement that the claims or theory have indeed been falsified. It is the obdurate refusal of their supporters to accept clear falsification that keeps creation science and Velikovsky's ideas alive.

Compare the tabloid science examples I have set out, to which can be added AD-X2, Velikovsky, and probably Krebiozen. Langmuir would probably have considered cold fusion, polywater, and memory water as additional examples of pathological science but would probably not have so classified the fifth force, although its rise and decline are typical.

In most respects, N-rays, cold fusion, and polywater are nearly perfect examples of Langmuir's pathological science: The effects were marginal, the implications certainly contrary to experience (and to well-established theories), and the rise and fall in the number of supporters followed Langmuir's script (though Benveniste's water never attracted enough support for its decline to be notable). There have been differences among these cases. In 1904 there was no good reason to be suspicious of Blondlot's claim to have discovered a new type of radiation, for the discoveries of radioactivity and X-rays had shown that a whole new field was only just emerging, and there were no guideposts. Today such a claim would be received with more skepticism, and indeed within the past twenty years there have been numerous "discoveries" that did not gain the notoriety that N-rays did.

In contrast, cold fusion and polywater attracted far more attention. Polywater survived for several years partly because the contaminants were harder to pin down, partly because of support from some hasty theoretical work. Cold fusion was skeptically received from the start but was promoted with a ferocious tenacity. Cold fusion was unique among these examples in its commercial pretensions, which served as an excuse for less than full disclosure. The fifth force differed from these in that there had been theoretical speculation for some time regarding the possibility of departures from the gravitational theory set out by Einstein. This remains a valid though not heavily populated area of research, and I would not classify the fifth force among the pathological cases.

ESP comes close to qualifying as a pathological science, for it meets many of Langmuir's criteria, but ESP is something of a mixture. Low-level ESP claims are invariably distinguished by fraud, poorly constructed research, or uncritical acceptance. What might be termed high-level ESP research does not, I feel, qualify as pseudoscience, though it has yet to establish itself as accepted science. It shares with Langmuir's pathological science the property of being based on very weak signals, but theorizing is not its strong suit, and the breadth of its support has not changed noticeably over the years. Still, it continues to attract the serious attention of a few competent scientists.

We can try to identify various cases with different parts of the fringe area. Pseudoscience might be the correct appellation for UFOs, for astrology and Jupiter effects, and for Lysenko's work. Creation science and Velikovsky's writings provide even better examples of pseudoscience. The fraudulent episodes in the medical sciences are also pseudoscientific, as they claim to be scientific yet do not contain results that are later confirmed. The cases of Millikan and Burt do not fall neatly into any tidy pattern, as they are neither pathological nor pseudoscientific. Burt probably did commit some fraud, and his work is now badly tainted, but it is not clear that all of his results are wrong. Millikan's case is troubling. With the notebooks for the critical year 1909 missing, we do not know whether he made any revealing comments. Knowing how many scientists operate, we can guess that he was guided by his instinct, but it would have been better if in his published papers he had given some discussion of his rejection of some results.

What we are seeing, pretty much as Einstein implied, is that classification may be easy in hindsight. It is the same with horse races, lottery tickets, and economic forecasts. Often, though, we cannot keep waiting for an uncertain vindication.

What we really want is a guide to help us evaluate the next "discovery," so that when it arrives we can quickly decide whether to ignore it, wait eagerly for the next update, or undertake some tests for ourselves. I like the way in which Hyman posed the problem. Suppose that

> A competent and respected colleague reports to you that he held a seance in his own house. During the course of the seance, one of the sitters asked if the medium could materialize a sunflower. Following this request, a sunflower six feet high fell upon the table. Your colleague produces affidavits from witnesses each of whom is a respected and honorable individual. He concludes that the only explanation is that the medium had access to a new force. ... Consider what your response might be.[10]

Hyman then went on to present his own analysis and also the comments he received from about a dozen invited respondents. Some agreed with Hyman, who took a very open-minded approach and in particular advocated thoughtful rather than knee-jerk criticism. Several respondents took issue with his analysis or what they read into his assumptions, and the philosopher Paul Feyerabend's initial reaction was, "Who cares?"[11]

Hyman is a psychologist who has taken a deep and critical interest in paranormal phenomena and ESP studies for many years. His sunflower question deals immediately with a problem in those areas, but it does not necessarily translate to the other types of fringe science that I have chosen to survey. My initial reaction to his question and to *all* paranormal claims of this type is disbelief. There have been far too many spectacular demonstrations that have turned out to be manifestations of our willingness, even eagerness, to be deceived under the guise of being open-minded. The best insurance against being fooled again is to insist on the presence of a professional magician, as Randi has always advocated. Scientists and especially physicists all too often suffer from the delusion that they are sufficiently hard-nosed skeptics, whereas they have too frequently turned out to be easy prey for slick con artists. But such demonstrations of "supernormal" powers will continue to occur, and most magicians are unlikely to want to waste their time as vigilantes. Randi's challenge is an interesting tactic: a prize of $10,000 for the first demonstration that convinces him. There have been few challenges, and none has succeeded. But Randi has placed the burden of proof exactly where it should be—on the person making the claim.[12]

What if we go beyond the stage of materializing sunflowers to the more carefully controlled studies of remote viewing, clairvoyance, and telepathy? In these the evidence is qualitatively different, though even the best of the studies appear to show only marginal effects, as Langmuir noted. Here we need to be careful. It is often claimed that the results of some extensive test depart significantly (in the statistical sense) from what one would expect on the basis of pure chance. It may well be that in the narrow sense this statistical claim is correct, but the validity of a claim of significance rests on two premises: (1) that there is no serious systematic bias that could easily overwhelm all other effects and (2) that the choice of statistical test is the most appropriate one.

Sources of experimental bias are notoriously difficult to locate. They can creep into the experimental design, into the conduct of the experiment, and into the later analysis. They can shift the measurements to values that are systematically higher or lower than they should be, or they can narrow or broaden the spread of measured values. Thus, when Jahn reports that his subjects could systematically achieve higher or lower scores, the effect has been small.[13] For example, in 262,650 trials in which operators attempted to score above the planned value of 100.00, the average score was 100.037. This result was strongly weighted by two operators who together had 98,400 trials. Among the other thirty-one operators, ten actually scored below 100.00 when trying for above average. An overall score of 99.966 was obtained when trying to score below average. Among the total thirty-three operators, fifteen scored above average while trying for a lower value, with no demonstration of enhanced powers that stands out far beyond statistical pleading. The overall distributions of scores do show small shifts in the averages that Jahn claims to be statistically significant, but it is notable that there seem to be no individual scores far from the average. What are we to make of these results and those from many other trials? This is where the intrusion of systematic effects can be serious and their exclusion so hard to establish.

The claim for statistical significance is also open to challenge. In their article "Statistical Significance and the Illusion of Objectivity," statisticians James Berger and Donald Berry have pointed out that the level of significance can depend critically on the type of statistical test applied.[14] Statisticians can easily disagree on the type of test to use.

As with so many claims for the discovery of some effect that lies outside the boundary of accepted science, Jahn's results do not yet compel acceptance of the existence of a paranormal effect. If there is a real effect, the burden of proof for its existence has yet to be discharged.

Claims for the discovery or demonstration of paranormal phenomena differ from other claims for scientific discovery, and organizations have been formed to actively pursue claims for demonstration of paranormal phenomena. The origins of the Committee for the Scientific Investigation of Claims of the Paranormal date to the 1976 "Objections to Astrology." But many of the organizers felt that their targets should be broader than astrology and that an ongoing campaign to alert the public was needed. The CSICOP has special committees on astrology, paranormal health claims, parapsychology, UFOs, and legal and consumer protection. The group's quarterly, *Skeptical Inquirer,* can be slightly shrill, but it is essential reading for anyone interested in this phenomenon. Its associated publisher, Prometheus Books, has a fine list that covers the bestiary of the pseudosciences.

Other groups of skeptics have formed around the United States and in many foreign countries (including the former Soviet Union), independent of but drawing their inspiration from CSICOP. A directory of these groups is included in each issue of the *Skeptical Inquirer;* their members include many knowledgeable specialists who will serve as speakers and as information resources for the local media. An organization with similar goals is the Society for Scientific Exploration. Its membership is drawn from practitioners of science who have acquired a firsthand knowledge and understanding of the scientific process. Though the SSE is concerned with careful scientific checking of claims for paranormal observations, it appears to be somewhat more open to sympathetic evaluation of paranormal claims, and it is not as active as CSICOP; it publishes the *Journal of Scientific Exploration* (with refereed papers). The efforts of the CSICOP and SSE have been directed mainly toward claims that relate to the paranormal and not to those that are routinely adjudicated within the boundaries of mainstream science; fifth forces, cold fusion, polywater, and memory water might attract a note in the *Skeptical Inquirer* but not a subcommittee.

These organizations represent a serious attempt to cope with claims that cannot be dismissed out of hand. The general public, still less the believers or near believers, will not be persuaded to disregard the latest "discovery" if scientists ignore it or think they can demolish it with a cute one-liner. Rather, the public will be confirmed in their view of science as a closed and rigid society, hostile to criticism or novel ideas. In the next chapter I have more to say about the need for the scientific community to have an evangelical arm.

Though many of the paranormal claims are patently trivial, modern parapsychological studies have reached the stage of sophistication where their critical analysis needs experts such as composed the National Academy panel. I find the verdict in the academy report well balanced: Nothing proven, but keep an open mind because of the integrity of the best of the current investigators. Still, skepticism is crucial, for favorable testimonials are not a substitute for scientific evidence.

Hyman's sunflower question was designed to pose the problem of evaluating claims for the observation of paranormal phenomena where scientific skepticism is strong. We can modify his question in order to refer to a discovery reported in a mainstream journal where peer review is routine and there is the usual expectation of acceptance of the published results. Suppose a long and detailed paper appears in a professional journal. In it we read of the discovery of (take your choice) anomalous water, cold fusion, a new force in nature. This discovery appears to invoke the existence of a process or substance not only previously unknown but perhaps contradictory to a well-established theory, inconsistent with a great many experiments. What is your reaction? Most scientists will be strongly influenced by the reputation of the journal where the report appeared. If it turned up in the *Journal of the Martian Academy of Sciences,* there will probably be little interest. It is a fact of scientific life that there are journals of low prestige, their contents generally ignored. Most of their papers, though probably correct, attract little notice apart from that of specialists and the abstracting journals. But if the discovery is reported through one of the major journals, it is likely that some experts will comment or undertake some research as a check or even realize that they have long had similar results that they had hesitated to publish. Where the authors have no track record (being listed, as a friend has phrased it, in the *Who's He?* rather than *Who's Who*), the "discovery" may simply be ignored. A similar reception or an educated wariness may greet the publication of results from scientists who have respected affiliations but are well known for the extravagance of their claims.

To a considerable extent our reactions also depend on our general knowledge of the subject or its related areas. With experience many of us develop an experience, an intuition, a gut feeling about how the world can behave. Thus, although there were widespread attempts to check the claims of Pons and Fleischmann, the major response within the physics community was one of disbelief. Until now we have not tried to quantify our intuition, and it is difficult to convey to nonscientists, to whom it can easily appear as a deep-seated conservatism or bias. Allan Franklin is a physicist who has tried to come to grips with this intuitive evaluation. He has written perceptively on scientific methodology, attempting to construct a model for this use of our prior experience.[15] He has drawn on a statistical technique known as Bayesian analysis and applied it to specific cases in particle and nuclear physics to make a very plausible case that we scientists are actually being guided by the accumulation of our knowledge, reasonably and systematically and more explicitly than we have realized. Scientists would probably be surprised to learn that so basic an activity as accumulating data should need justification, but Franklin gave the title *The Neglect of Experiment*[16] to his first book to draw attention to this omission, and he cites studies in the current philosophical literature that give greater attention to the role of experiments.[17]

It is professional intuition that lies behind the instinctive professional rejection of claims such as those of Velikovsky, of Geller, of astrologers, of those who try to refute relativity. We do not go to the trouble of making calculations in each case to show where Velikovsky or antirelativists are wrong; our experience tells us that this is a waste of time. There is, of course, the danger that in this way we may be refusing to accept a true discovery, but the record shows this to be rare (though the exceptions can be prominent). Yet we do have to balance the competing demands on our time and do not want to be endlessly distracted. At least most of us are better placed to make these intuitive judgments than are the many people who make or support the contested claims.

Claims for scientific discovery require that we get evaluations from people knowledgeable in the appropriate fields. Those of us who are not similarly expert can either defer to the experts or try to educate ourselves. Where expert opinion is overwhelming, that may in itself be telling us something, but when there is dissent from a significant and respected minority, we have a strong indication that more data are needed.

In mature fields a tentative approval by experts is indicated by publication in a professional journal, implying the possession of a minimal degree of credibility to each new report or claim. This is not a leak-proof warranty but rather an invitation to take the report seriously. At this point we have a contradiction. The professional parapsychological journals, even though refereed, do not command sufficient respect among mainstream scientists for their contents to be automatically accepted. I think that I am safe in saying that the vast majority of scientists would not be persuaded by a paper in the *Journal of Parapsychological Research*. Conversely, there is no parallel and generic mistrust of papers that appear in *Nature*, *Science*, or the *Journal of the American Medical Association*.

There are other components of fringe science that I have not yet covered in this roundup of reactions to novelty. For example, what happens when a respected scientist holds tenaciously to views that are opposed by the vast majority of the experts? There are many examples of just such cases. Pauling, Nobel Prize laureate in chemistry, has long been an advocate of the effectiveness of vitamin C in preventing colds and cancer. The medical profession sniffs at the idea, and clinical trials have not demonstrated the effectiveness of vitamin C. Fred Hoyle is a theoretical astrophysicist of great ingenuity and repute. For some years, he has been fighting a lonely battle on behalf of the idea of panspermia, the theory that life did not start spontaneously on earth but came here from outside. With a few colleagues he continues to publish in mainstream journals, analyzing the infrared spectra of comets and interplanetary material and arguing that there are signs of the organic molecules that are needed to generate life forms. Most other scientists remain unconvinced. A similar isolation affects Halton Arp, who over many years carried out highly regarded observations of quasars with the 200-inch telescope on Mount Palomar. Arp has produced evidence that he claims disagrees with the distances generally assigned to some quasars. The scientific debate is important, for distance measurements are critical to cosmological theories. A very few astronomers support Arp. After several years Arp was denied further observing time on the Palomar telescope. The reason given for this exclusion was that he had not produced anything more conclusive, that time allocation on the big telescope was at a premium and he had already been allowed sufficient time. This is a form of censorship but not unreasonable under the circumstances.

Another example of a scientist's exhausting the patience of his colleagues is that of Peter Duesberg, who has an established reputation as a molecular virologist. He has long campaigned against the generally accepted view that HIV infection is the cause of AIDS. My medical friends tell me that he has come up with provocative ideas that could not be promptly ruled out, but that careful checking has so far failed to support any of these ideas. In attempting to gain a wider readership for his views, he submitted a letter to *Nature* criticizing a paper that had reported a significant increase in HIV infection among heavy drug users. John Maddox, the editor, not only declined to publish it, but in a strongly worded editorial, was highly critical of Duesberg's persistent holding to views that others have shown to be in error. The editorial concluded with the remark that "when he offers a text for publication that can be authenticated, it will, if possible, be published—not least in the hope and expectation that his next offering will be an admission of recent error."[18]

Recognizing that the referee system of professional journals also serves as censor, the American Physical Society (APS) has devised a way of giving a platform to the truly unorthodox views, those that will not meet the standards of its journals (views that are often not even submitted for publication). The APS has several meetings each year, held in different cities. There are many sessions for "contributed papers," accepted from any member of the society. An abstract of every paper appears in the meeting program, published in the *Bulletin* of the APS. The author of each paper has an allotted time at the meeting. Papers are grouped into sessions by topic, and most meetings have one or two sessions filled with a miscellany of papers. In them you find the eccentric and the unorthodox as well as the unfashionable and the isolated. There is, of course, no compulsion for anyone to listen to these presentations. The only requirement for acceptance of a paper for a meeting is society membership. Because of the wide distribution of the *Bulletin,* this system promotes some awareness of the existence of the unorthodox ideas. Anyone so alerted and so inclined is free to write to the author for a copy of the manuscript, but this does not carry even the limited warranty that is implicit in full and refereed publication.

But what of nonmembers or would-be contributors in disciplines that do not provide this safety valve? Most true cranks do not generally try to insert themselves into mainstream science. Some (at least in physics) mail out copies of their manuscripts; a few even pay for the publication of books that they distribute. The unsolicited manuscripts with denunciations of relativity or whatever are best left unanswered; experience shows that even a courteous response soon leads to a fruitless exchange.

As has been noted before, it is not only the pseudoscientists who have a problem in gaining the attention of the established scientists. In an oft-cited paper, "Resistance by Scientists to Scientific Discovery," Bernard Barber gave many examples of this resistance: Lord Kelvin, one of the giants of nineteenth-century physics, could not bring himself to accept the findings that atoms were not indivisible entities; Planck, originator of quantum ideas, noted that "a new scientific truth does not triumph by convincing its opponents and making them see the light, but rather because its opponents eventually die, and a new generation grows up that is familiar with it."[19] (This is perhaps an extreme view, but it has a grain of truth.) Einstein was never philosophically satisfied with the form of quantum theory and its predictions, despite its clear successes and his unsuccessful attempts to argue against it. Barber identified several reasons for such reluctance. One is that there are genuine grounds, that there is good scientific argument against the innovation. Sometimes the innovation is phrased in terms of novel or unfamiliar language or mathematics, as was the case with relativity. Sometimes there are religious objections, as with evolution. Perhaps the new ideas come from "outsiders," and the objections (valid or not) are obvious to the "insiders" (consider continental drift and cold fusion).

This is not an exhaustive list. On the whole, though, and understandably, mainstream science is necessarily conservative. The most efficient way that science can make progress is by avoiding endless diversions. There is a barrier of plausibility that must be surmounted before a claimed discovery will be taken seriously. Reading the latest manuscript or a published paper, we are aware that many scientists are known for the reliability of their work, others (sadly) for their unevenness.

To this point I have confined my attention to the responses of scientists. What are nonscientists to make of the various claims? "Trust the experts" is probably best in most cases, but experts can disagree in their evaluations even if there are no social or political overtones. The disagreements over scientific matters that involve public policy can often be distilled into differences in the scientists' political or social preferences and should be seen as such. For example, there may be agreement that there are insufficient data to decide whether some particular level of a chemical is safe in foods. The disagreement then centers on the appropriate policy to adopt in the absence of additional data. At what level should the FDA require remedial action? This is the sort of problem of great practical importance that turns up again and again in environmental issues and will not go away; a regulatory agency will be pressed either to leave the presently permitted levels alone (and anger one lobby) or tighten them (and anger others in a different way).

In many places I have talked about the "experts" (see, for example, Chapter 5). Just who qualifies as an "expert"? Expertise or authority do not automatically accompany a doctoral degree. Colleagues judge expertise on the basis of the accumulation of respected professional publications. A graduate degree usually helps but is not essential. Some scientists become widely accepted as authorities because of the importance and continuity of their scientific contributions.

Unfortunately, the title of doctor or professor is often brandished to impress the nonscientist when some novelty is being presented, usually through nonprofessional channels. Expertise does not automatically extend beyond one's original field but can be demonstrated. Dr. Andrew Ivey was clearly out of his professional depth when he was being used to endorse Krebiozen, but the prestige attached to his university position and his career were exploited in the efforts to market that bogus cancer cure. (This is no different from the endless endorsements of all sorts of products by athletes and other celebrities, but we might have hoped for more from a scientist. Ivey was sincere but misguided.) Pons and Fleischmann were respected electrochemists but soon demonstrated their lack of expertise in the detection of fusion products.

As we wander further from the mainstream, we find greater emphasis being placed on the opinions of people who, quite simply put, do not know what they are talking about. (We might add this as a criterion for identifying the crank or pseudoscientist.) Velikovsky's *Worlds in Collision* received rapturous endorsements from people who were, to be charitable, ignorant and confused about some very basic principles of physics. Testimonials, in fact, are the stock in trade of the pseudoscientists, from the AD-X2 case through the fake cancer cures to a range of paranormal phenomena. Testimonials of this sort carry no weight in a scientific disagreement, but they can seem impressive to a public that often mistrusts experts, empathizes with other nonexperts, and has a deeply felt longing for medical miracles.

We return to the central question and must accept that there is no simple and comprehensive test that can be applied in the classification of a new idea. Mainstream science is fallible but generally correct, and its conservative approach constitutes a reasonable response. As in sports, the actions of the participants determine the outcome; the crowd, no matter how vociferous, does not score a touchdown or a home run. New science must therefore prove itself by its contents and not by its advertising or testimonials.

Notes

1. See, for example, Tristam Englehart and Arthur Caplan, *Scientific Controversies* (Cambridge: Cambridge University Press, 1987).

2. Cited by J. W. Grove, *Minerva* 23 (1985): 216.

3. *Science* 141 (1963): 1010.

4. Ibid., 1012.

5. Ibid.

6. *Physics Today* 42 (October 1989): 36.

7. Ibid.

8. *Skeptical Inquirer* 9, 1 (1984): 36.

9. Martin Gardner, *Fads and Fallacies in the Name of Science* (New York: Dover, 1957), 11; previously published as *In the Name of Science*.

10. Ray Hyman, *The Elusive Quarry* (Buffalo, N.Y.: Prometheus Books, 1989), 243.

11. Ibid.,

12. James Randi, *Flim-Flam! The Truth About Unicorns, Parapsychology, and Other Delusions* (New York: Lippincott and Crowell, 1980).

13. Robert G. Jahn and Brenda J. Dunne, *Margins of Reality* (New York: Harcourt, Brace, Jovanovich, 1987), 352, table A.

14. *American Scientist* 76 (March-April 1988): 159.

15. *Experiment, Right or Wrong?* (Cambridge: Cambridge University Press, 1990).

16. *The Neglect of Experiment* (Cambridge: Cambridge University Press, 1986).

17. *Experiment, Right or Wrong?* 1.

18. *Nature* 363 (1993): 109.

19. *Science* 134 (1961): 596, 597.

13

Responding to Fringe Science

Mainstream science must have its fringe component if it is not to atrophy. In this fringe there are discoveries that will later graduate into the ranks of the accepted and others that will never receive confirmation. How scientists will view novelty will depend on its immediate degree of plausibility. We can certainly anticipate the arrival of other cases of fringe science, if not specifically then generically. The purpose and style of this anticipation will take various forms, depending on the particular fringe area, and scientists will respond differently from nonscientists.

As long as the general public does not or cannot distinguish between science and its imitators, it will risk rejecting scientific advice or falling victim to the many unscrupulous operators who are willing to exploit the gullible and the desperate. But here again, attempted remedies must be based both on an identification of the conditions that provide a foothold for pseudoscience and on an acceptance of the permanence of a phenomenon that is reflective of human nature.

The problem is not new. In his 1852 compilation "Extraordinary Popular Delusions and the Madness of Crowds," Charles Mackay documented the mass deceptions that lead to financial catastrophes such as the eighteenth-century South Sea Bubble. He denounced "alchymists ... fortune-tellers and astrologers ... [and] almanac makers." "We find," Mackay wrote, "whole communities suddenly fix their minds upon one subject, and go mad in its pursuit; that millions of people become simultaneously impressed with one delusion. ... Some delusions, though notorious to all the world, have subsisted for ages ... for instance ... the belief in omens and divinations of the future, which seem to defy the progress of knowledge to eradicate them entirely from the popular mind."[1]

On a related topic, there was a long discourse in the *Fortnightly Review* of June 1873, "On the Dread and Dislike of Science," that noted that science "is so little appreciated by the world at large that even men of culture may still be found who boast of their indifference to it."[2] Nearly a century later the extent and durability of this ignorance of the content and methods of science was the centerpiece of C. P. Snow's 1959 Rede Lecture, "The Two Cultures."[3] To be considered as cultured and educated, Snow complained, it was sufficient to be well versed in the classics of literature though totally ignorant of the second law of thermodynamics. The nonscientists' eager acceptance and endorsement of Velikovsky's ideas and other fringe science makes the same point repeatedly. One could compile a large volume documenting the demonstrations of this educational and cultural asymmetry.

Scientific fraud is in some ways easier to handle, at least procedurally, for this is initially a problem internal to science. Not only is the reputation of science at stake, but so is the delicate system of mutual trust on which science so critically depends. The remedy for fraud is in the hands of the scientific community. This requires vigilance on the part of journal referees and a greater sharing of responsibilities in multiauthor papers. The Baltimore case drew attention to this latter aspect of modern science that can lead to errors or omissions. Forefront science is inherently risky. Techniques are often being developed as the research progresses, and the range of required skills often exceeds that of any single scientist in modern collaborative research. Collaboration is therefore inevitable and a source of strength, but with it comes a division of scientific responsibility. Cross-checking, even of a final manuscript, may be minimal because of the intellectual separation of the cooperating disciplines. There can be no universal rule governing how much every author should know about every paragraph in a final manuscript; collaborators need to trust one another.

None of these steps will be fail-safe. Even more important and especially in the medical sciences is the need to remove the incentive for cheating or the rewards for carelessness. The extreme competition in the hunt for research funds and for gaining promotion has induced the least scrupulous to stoop to the sorts of dishonesty I described in Chapter 10. In response some university promotion committees and some of the federal agencies that support research have stated that they will consider no more than about five publications in the supporting documentation for any grant application or résumé. This approach can remove some of the reward for accumulating enormous numbers of publications. It can also help to reduce the imperial and widespread practice of "honorary" authorships, in which senior scientists insist on being listed among the coauthors of all papers emerging from their laboratories, no matter how tenuous have been their contributions to the work.

An added source of pressure to keep science to the straight and narrow has come from the National Science Foundation (N.S.F.) and National Institutes of Health. As a condition to receiving research funds, every university is now required to establish procedures for responding to allegations of academic dishonesty. There need to be strict safeguards for maintaining confidentiality during investigations because unfounded allegations can leave permanent stains even if proven false. Probably as important as any of the statutory steps will be the climate of increased ethical awareness that they generate. The scientific community needs to be seen to be policing itself in order to ensure acceptance of its findings and continued funding for its research. No good will come from a serious erosion of public trust in science. The record of recent years has shown that in-house investigations can be less thorough than needed. But it is a giant step from admitting laxity in reviewing allegations to bringing in the Secret Service. There is a chilling effect in the specter of faceless people in trench coats looking into lab notebooks and analyzing the ink and paper.

A second problem also largely internal to science centers on the peer review system. Disgruntled and rejected authors have often (sometimes with justification) complained about bias, and abuses have been alleged in the reviewing system for research grants. This has brought an important and deplorable response, the growth of what is termed pork-barrel science. Institutions of renown as well as ambitious but undistinguished institutions have besieged members of Congress from their home states, often using high-priced lobbyists, to gain access to the federal trough and in this way bypass peer review. The quality of the proposed pork-barrel science is often depressingly low, and the existence of this style of patronage is corrosive to whatever quality control peer review is designed to provide. Attacks on the peer review system undermine the trust in science and its objectivity, which are our guarantors of reliability. Concern for the overall integrity of science should provide a motivation for the scientific community to be more vigorous in its efforts to police itself and improve the operation of the peer review system.

In the critiques of the departures from honesty that are paraded before us, there is a mixture of cases—the well known, the obscure, and the trivial—discoveries initially rejected that gained only delayed and reluctant recognition, if any. Among the examples cited are many that are mainly of historical interest, with little if any current relevance. Conditions in science have changed almost totally since Planck had difficulty in getting his ideas on thermodynamics taken seriously. In those days the structure of science was far more hierarchical than it is today. *The* professor was in charge, in omission and commission. Junior scientists had virtually no alternative avenues open, so dependent were they on their professors for advancement. Today there are many professors and far more journals (perhaps too many) in which to publish. It is notable that the remnants of this historical academic feudalism are strongest in the medical sciences, and that is just where a disproportionate number of abuses seem to occur.

It seems that science is far from its proclaimed dispassionate objectivity, far from trustworthy. There have indeed been well-documented instances of fraud, personal bias, as well as innocent error, but I think the general picture of science is often veiled by a condemnation of the entire system. Omitted is the recognition that the total number of these cases is remarkably small and that the overall efficiency and success of the system are impressive.

The greatest challenge to our ingenuity and staying power comes from the outrageous pseudoscience that regularly emerges like mushrooms to receive uncritical promotion by the media, directed to a nonscientific public that is both receptive and unequipped to evaluate each new claim. Thus, as we turn from science to the pseudosciences, we find different problems that require different solutions. A yearning to believe or a generalized suspicion of experts is a very potent incentive to accepting some pseudoscientific claims. It is easy to prey on ignorance and fears while suggesting that cellular phones can cause brain cancer. The promoters of Krebiozen, Laetrile, and other "cures" prey on the desperation and vulnerability of cancer patients and their families, for whom conventional medicine can offer no hope. This exploitation of the mantle of science is unique as the most cruel and despicable of all of the fringes we have encountered. In comparison the widespread acceptance of astrological horoscopes might have been thought to be benign and even amusing, until it was found to have penetrated to the highest office in the land. One might have hoped that in a modern industrialized country such superstition would be a thing of the past. Many people read the astrological columns in the daily newspapers, mostly for amusement but probably also out of curiosity, half hoping, half fearing. Astrology might seem like a harmless addiction, and it probably would be were it not a part of a much larger picture—the general lack of scientific knowledge within an often ambivalent acceptance of both science and pseudoscience.

The popular receptiveness for astrology and the occult has been studied by Barry Singer and Victor Benassi, among others. They see it as a "challenge to the validity of science and to the authority of the scientific community ... probably based on psychology and sociology rather than rational determinants."[4] To some extent they agree that many people hold mutually inconsistent beliefs in well-insulated compartments of their minds, but they also point to underlying deficiencies in science education, where "science seldom seems to be taught as a cognitive tool."[5]

Those of us who teach science survey courses have seen this regularly. Students can "learn" the material and pass our exams, but we have not been particularly successful as advocates of scientific rationality. Students are rarely exposed to any except the most superficial discussion of scientific methods, and, not surprisingly, they emerge from our classes with a picture of science that is unbalanced through omission. Even their factual knowledge fades rapidly after the final exam. Why should we hope for greater success in persuasion when we confront pseudoscientific beliefs that are much more firmly and personally held? And if we have so little prospect for success, why should we keep on trying?

The immediate answer has to be that the alternative is less attractive. We might change the question to asking how much worse might the situation be if we were not trying to combat scientific ignorance. We must, I believe, continue our efforts to improve science education and our missionary work, hoping for whatever converts we can make and hoping that some of this will help to shape the public perception of science and reduce the influence of the pseudosciences.

What's to be done? Leading all other efforts must be the extension and improvement of science education. Singer and Benassi include this among their recommendations, urging that science be "more easily apprehended as a highly useful mode of inquiry."[6] Present curricular invigoration has as a major objective the recruitment of more young people into scientific and technical programs. This is a worthy objective, but it must also confront the vast majority of students who will not go into scientific careers. It has long been recognized that a campaign of this kind cannot start with high school science but needs to penetrate to the lower grades, since attitudes toward science, even toward learning, are established at early ages.

Changes as widespread as are needed will come slowly. Until then, as those curricular reforms are slowly introduced, we must not ignore the students who have already graduated from our schools and are now in our colleges. Just about every college requires its students to take some science courses, even if these are far from their major area of concentration. In this way scientists have an opportunity, through survey courses, to reach most of the people who will be influential in coming years. We can shape their impressions and add to their understanding of science, its structure and its content. This is an opportunity that has been sadly underused. Most of the present science courses are heavily laden with facts. Each of our scientific disciplines has so many facts that we think are interesting and important that most instructors have difficulty in deciding which topics to leave out. The results can be seen in the encyclopedic textbooks. "Facts" are easy to test students on, and the temptation to do so, with multiple-choice questions, becomes irresistible when an instructor faces a class of many hundreds. What is missing from the parade of facts, laws, and the latest discoveries is scientific method. Many texts do give a fleeting look at "the scientific method," which, along with other roadkill, recedes in the rearview mirror as the semester speeds along. What is needed is continued reference to method, not through a heavy-handed and abstract introduction to the philosophy of science but through continually asking the questions, How do we know this? How was this discovered? We should provide illustrations as well.

We need to show the rationality of science: how planned experiments are carried out, how the results are synthesized into overarching theories. We need to point to the limitations of our knowledge, the tentative nature of some conclusions and the greater confidence we have in others. At present our curricula are too often recitals of unqualified successes. We should not omit the accidental discoveries or the superseded theories, for we can use these to show how they, too, have been logically fitted into an evolving total structure. We can and should introduce some current pseudoscientific episode and draw on the methods we have developed to answer the question, How do we deal with this new claim? Such comparisons can be entertaining as well as educational. This sort of foundation provides a framework against which students can measure future pseudoscientific claims.

Most pseudoscience comes from outside the mainstream of science. Confused in their understanding of science, deeply entrenched in their views and unswerving in their adherence to their "discoveries," the pseudoscientists would be of no consequence if they were not supported, encouraged, and used by the media. What is depressing is the frequency and the enthusiasm with which these people, so pitifully out of their depth, are paraded before us. The episodes slowly fade, but in the meantime this nonsense masquerading as science attracts viewers and readers.

We are repeatedly regaled by stories of psychics helping police to locate missing people, of remarkable coincidences that are imputed to demonstrate some supernatural something, of quacks with their nostrums and inventors with their perpetual motion machines. There has been a steady stream of similarly readable nonsense—on pyramid power, on sentient plants, on crystals, on Jupiter effects. Do the publishers and radio and TV show producers bear any responsibility for the accuracy of what they produce? It is hard to avoid the inference that sensation is more important to them than accuracy.

The scientific community is rightly expected to hold to the highest standards of honesty. We used to think that this was routine, completely natural, but experience has shown otherwise. There is now an increased awareness of the potential problems. The NIH has an Office of Research Integrity. There has even been a congressional hearing that focused on the allegation of dishonesty in a scientific paper. Why should not pseudoscience and its promoters be held to an equal standard of honesty just because they have not received public funds? When a pseudoscientific book based on sloppy science (deliberate or otherwise) is put out as though it were the greatest scientific advance since sliced bread, is there no shame? A major publisher is pushing a series of books on psychic drivel, huckstering them on national TV. Does truth in advertising not extend to the claims of publishers? The short answer to these questions is Caveat emptor. What, then, should be the responses of the scientists?

It is easiest for scientists to ignore the pseudoscientists. Most do. We cannot expect every scientist to take issue with every pretender, for there are more interesting and more productive things to do. By the same token, it would not be acceptable for every scientist to turn away. Unless there is expert and cogent criticism, the outlandish claims will, by default, come to be accepted by nonscientists, even by scientists in other areas. Strident denunciations will not suffice, indeed will be counterproductive. Some positive action is needed, and CSICOP and SSE provide both a model and resources. Singer and Benassi urge the "issuing of rebuttals to fraudulent or unsubstantiated occult claims," and they consider this "a compelling social responsibility."[7]

In the never-ending campaign against the often attractive or plausible pseudosciences, there are several tactics or precautions that scientists should employ:

- Be 100 percent accurate in your criticism. Far too often an indignant scientist will make errors that then become the focus for further debate, diverting attention from the original subject. It is better to be completely correct than to have a great rejoinder but be even slightly in error.

- Do not trust the accuracy of citations to the scientific literature when they come from a likely pseudoscientist. These citations are designed to give the appearance of expert support. Check them out. They are often incomplete, selective, or distorted.
- Do not overstate the case for orthodox science. Be prepared to admit the open-endedness of science, the gaps yet to be filled even while the main structure is widely accepted (as is the case with evolution).
- Do not accept the simplistic argument that because it took so long for Wegener's idea of continental drift to be accepted, therefore the current claim (whatever it is) must be correct and will also come to be accepted, even though the experts now oppose it. Know the facts of the Wegener case and be prepared to correct its misuse. This particular example is a favorite of pseudoscientists. Another example of scientists' error relates to the ability of bees to fly. It is often stated that a theoretical physicist calculated that bees could not fly. As far as it goes, this is correct, but an important detail is usually omitted from this example: The physicist started by assuming that bees' wings were rigid. Faced with the theoretical conclusion that bees could not fly and the reality of actual flight, he immediately realized that the wings had to be flexible, changing shape during flight.
- Be able to cite counterexamples of the willingness of science to entertain radical proposals that might have scientific merit.
- When confronted by testimonials, probe their reliability. Were there independent witnesses? Were there repetitions? Were there controls?
- In writing and in face-to-face debates with pseudoscientists, be prepared to confront them with examples of their errors. This must be done with precision and by explicit citation. In a debate have at hand photocopies of relevant documents. There is nothing like this to block a pseudoscientist's evasions (as I found in my encounter with Velikovsky). A colleague who has debated with prominent creation scientists tells of his success with similar tactics. In olden days holding up a cross was sufficient to stop the devil; today photocopies can be effective exorcisers.

Having addressed the obligations of the scientific community and the needs of scientists who care to confront the pseudoscientists, I turn now to the problem that faces the nonscientist: What to do when the next pseudoscientific claim emerges. Just as importantly, what should be the response of journalists to the next pseudoscientific report? Skepticism is a great protection. The claim may sound plausible and interesting, but who is putting it forward? What credentials does this person have? This may not be easy to judge, as in many cases a title is displayed but its provenance omitted, and a nonspecialist might be unable to judge its relevance. Has this claim been tested through normal professional channels? If the answer to this last question is no, that is a clearly flashing danger signal. Have knowledgeable professionals supported the claim or criticized it? If the weight of expert opinion is strongly opposed to the new claim, be warned. The experts may not always be 100 percent right, but they are certainly not always 100 percent wrong. In many instances the novel claim does not need an immediate response. No tangible harm is done by delaying acceptance, to allow time for evaluation or debate between the experts and the innovator. For the journalist, hard pressed by rapidly approaching deadlines, tracking down expert opinion may be a disposable luxury, but haste carries risks.

At the end of our survey of fringe science, where do we stand? Are we any better placed to classify the next scientific claim that arrives? Although we have not solved the demarcation problem, we have pointed to its complexity. Neither science nor its fringes will wait for an agreed solution. In the scientific literature where peer review has been a hurdle initially surmounted, reports can be treated with initial respect, though a cautious skepticism is always useful. Outside the mainstream, professional experience is still the best guide for the evaluation of novelty. Our scientific nostrils should also allow us to detect the aroma of pseudoscience where, as has often been asserted, extraordinary claims demand extraordinary proofs.

For professionals in the media, skepticism is even more important. The breathless reporting of each new claim for some breakthrough should be restrained by the recollection of the long record of claims made prematurely and not later confirmed. Medical researchers are not blameless in the style of their announcements, and one must rely on the tempering that the journalists can provide. But even as science journalists work to protect the accuracy of their reporting, there is a continuous undermining of their efforts by the willingness of editors to report uncritically "successes" of psychics in finding missing persons or the pronouncements of public figures who (often knowingly) misinform their public. For example, when discredited ideas about the sources of atmospheric pollution are eagerly recounted by agenda-driven commentators or politicians, the public are the losers. The media need to expose this dishonesty as eagerly as they pursue other betrayals of the public trust.

There remains the nonscientific public, caught between, on the one side, the extreme claims by both scientists and pseudoscientists and, on the other, a vigorous skepticism that spills over into the total and undifferentiated disbelief in all scientific claims. There are no easy remedies. Perhaps to those of us professionally engaged in science, the criticism as well as the praise should be comforting, for these demonstrate the importance attached to science. After all, who would bother to criticize or imitate something of no value, let alone argue over its philosophical or social aspects?

Notes

1. Charles Mackay, *Memoirs of Extraordinary Popular Delusions and the Madness of Crowds* (London: National Illustrated Library, 1852).

2. *Fortnightly Review* 138 (June 1, 1878): 805.

3. The lectures were later published as *The Two Cultures and the Scientific Revolution* (Cambridge: Cambridge University Press, 1959).

4. *American Scientist* 69 (January-February 1981): 49.

5. Ibid., 55.

6. Ibid.

7. Ibid.

Epilogue — 1998

Since *At the Fringes of Science* appeared, no new and major categories of non-science have opened up. There has been progress, of sorts, in some of the cases that I had described, but it is clear, though not surprising, that there is still no widespread understanding of the scientific method and what it is that separates science from nonscience or pseudoscience. It was not my purpose, in writing *Fringes,* to try to settle what philosophers of science term the *delimitation problem* by defining precise and immutable boundaries for science, boundaries that those of us in the interior would defend to the death. Rather, it was to show that there are, in the neighborhood of widely accepted science, many different regions such as the almost-science, the not-yet-accepted-science, agenda-science, shading imperceptibly up to the region of the clearly cranky.

Who decides how each new claim for acceptance is classified? It has been my thesis that it is the scientific community that makes these determinations. There are determinations that later prove to be wrong, and the scientific community can, and does, change its collective mind, but what the scientific community accepts as the best science at any time comes only from its evaluation and not from elsewhere. However, although the coalescence of the "best scientific opinion," may be delayed by the scientific community, feeling that there was insufficient or only contradictory evidence, the outside world sometimes cannot wait. For example, substantial damages are being claimed in one lawsuit on the basis of the effects of secondhand tobacco smoke; and in another, the central question is the effect on a fetus of the mother having been given Bendectin, a medication designed to alleviate morning sickness and since withdrawn by its manufacturer. But the courts cannot wait for science to reach a firm conclusion.

In such cases, the contested issues usually hinge on product liability problems or on health effects, such as those alleged to stem from electric power lines or various medications. These subjects are far more technical than the judge or jury is able to evaluate unaided, and scientific expert witnesses are then called upon to testify, to explain the issues in everyday language. Who is qualified to claim the status of "expert witness" and what should be the nature of allowable testimony?

In a major decision in 1923, the U.S. Supreme Court (*Frye v. United States*) ruled that the test for admissibility of evidence was "general acceptance": "The thing from which the deduction is made must be sufficiently established to have gained general acceptance in the particular field in which it belongs." This was the controlling decision until 1993, when the Court (in *Daubert v. Merrell Dow*) unanimously set Frye aside and held that the Federal Rules of Evidence govern the admission of evidence. In delivering the Court's decision, Justice Harry Blackmun observed that "Rule 702 . . . clearly contemplates some degree of regulation of the subjects and theories about which an expert may testify. . . . The adjective 'scientific' implies a grounding in the methods and procedures of science." He went on to quote from Karl Popper's magisterial *Conjectures and Refutations* and physicist John Ziman's *Reliable Knowledge*. He also noted that "Rules of Evidence [are] designed not for the exhaustive search for cosmic understanding but for the particularized resolution of legal disputes." In December 1997 in its ruling in *General Electric Co. v. Joiner,* the Court clarified the Daubert ruling and strengthened the power of trial judges to filter out "junk science."[1]

The issue of the reality and reliability of scientific "facts" and theories is obviously central in legal disputes, but it has also emerged as the focus of an all-too-academic debate. The structure of accepted science includes data from experiments and observations and the theories that represent our current best attempts at creating an overall description of our world. Our theories also provide a framework for the testing of the validity of new claims that may emerge.

Most scientists will agree with me in considering science to have a base in physical reality. Scientists are, of course, embedded in their societies and have personal views and ambitions that are often reflective of social and political views of those societies. Nevertheless, our science, we believe, stands apart, with its content possessing some objective reality. In situations where the social and political preferences have influenced the content of the science itself, as happened with Lysenko in Russia, the results have been bad science of no lasting value but with very serious temporary damage. Nevertheless, there is at present a very different view of science, held most notably by some sociologists and philosophers of science. In this "relativist" view, probably not even known to the majority of scientists, there are no absolute scientific truths, but the claimed "truths" are at best reflective of the scientists' own social or philosophical or political views. This strange view has been rejected with varying degrees of scorn by most mainline scientists who have come across it.

The emptiness of the relativist view was exposed in 1996 by Alan Sokal, a theoretical physicist, whose paper "Transgressing the Boundaries: Toward a Transformative Hermeneutics of Quantum Gravity" was published by *Social Text*.[2] This paper was impressive—an apparently scholarly treatment of a difficult subject, replete with massive documentation in footnotes. It has been described by Barbara Epstein[3] as "an amalgam of postmodernism, poststructuralist theory, deconstruction and political moralism," each keyword being a panic button that would cause the knowledgeable to salivate. Unfortunately for *Social Text*, Sokal's paper was a hoax, designed to show how the editors (and presumably most of the readers) would be unable to recognize (as Sokal put it) "an article liberally salted with nonsense if (a) it sounded good and (b) it flattered the editors' ideological preconceptions." Sokal's essay included a torrent of trendy adjectives that disguised an intellectual vacuum; you can follow the continuing saga through Sokal's web page.[4] The debate leaves the scientific community largely untouched but constitutes a source of embarrassment, amusement, and/or indignation, depending on one's affiliation and worldview.

Very different from these skirmishes has been the Baltimore case, in which allegations of fraud in biological research were made against Thereza Imanishi-Kari by her junior coauthor, Margot O'Toole (see pp. 136–138). Their senior colleague, Nobel Prize winner David Baltimore, defended Imanishi-Kari with vigor and was tarnished for what was seen as an uncritical defense. This case centered around a paper published in 1986; it came to the public attention in 1989 when it was the focus of a volcanic congressional committee investigation. Since *Fringes* appeared, there has been yet one more review, which ended with a dramatic reversal of the earlier judgments and the exoneration of the principals, Thereza Imanishi-Kari and David Baltimore, though with criticisms of Imanishi-Kari's "sloppiness" in the laboratory.

This final review was carried out by a panel of three judges appointed by the Department of Health and Human Services (HHS); earlier reviews had been carried out at MIT and Tufts University, then by the Office of Scientific Integrity (OSI) and its successor, the Office of Research Integrity (ORI), these latter two under the NIH. The report of the HHS panel, issued in June 1996,[5] contained some very strong criticism of the OSI and ORI procedures.

What lessons can we draw from this unhappy tale? First, there can be honest differences in the interpretation of scientific data without any trace of fraud or dishonesty. Scientific disagreements are often robust. Scientific truth will generally emerge (if at all) only after replications of the experiments, the confirmation of extrapolations from the data or the consistency (or lack thereof) with yet other experiments—and the timescale for this resolution can be very long, as happened with the radical idea of continental drift. Second, the fairness of the judicial procedures for handling accusations is essential if the judgments are to be insulated from pressures—personal, scientific, or political—and command respect. In this regard, the involvement of the accuser (Margot O'Toole) with the subsequent investigations was simply too close. Accusers, quite naturally, seek vindication and must be distanced from the investigation.

Probably of greater importance in this case was the spectacle of a congressional hearing, making the Baltimore case unique among investigations where there have been allegations of scientific fraud. In no other country has a charge of scientific fraud led to anything like a congressional hearing, with its limited procedural protections. This U.S. political-judicial mixture is unique, and although it has been a powerful combination for illuminating illegal, corrupt, or borderline conduct in public affairs, it is clearly too blunt for resolving scientific cases. In the Baltimore case, resolution took more than ten years, with resulting damage to careers.[6]

The lack of understanding of the functioning of science and of the nature of validation of claims involving science also becomes of more than academic or legal interest when it involves the expenditure of over $10 million per year from public funds on an Office of Alternative Medicine. The scale of "alternative medicine" is very large. It has been estimated that more than 50 million people, within the United States alone, use these remedies or treatments each year, spending more than $10 billion. Some medical insurance companies are now willing to pay for claims for treatment by "alternative medicine."

What, precisely, is "alternative medicine"? Around the world, people have developed folk remedies and treatments for all sorts of ailments. In some cultures, the dispensation of these traditional medications has been in the hands of medicine men or priests, in others, the practices are part of the collective folk wisdom. The effectiveness of these treatments is accepted on the basis of anecdotal evidence, and that is at the center of the present debate in the United States. The spectrum of public opinion ranges from the view that "there must be something to it, otherwise it would not have been used for so many years" (sad evidence of public gullibility), to a total skepticism.

In principle, there may well be effective procedures or medications embedded in these folk practices, and if so, there should be formal validation (as, for example, by the FDA or NIH) and availability throughout our medical system. How, then, to conduct the needed tests? Machinery and well-defined rules for clinical tests already exist. Most medical research in the United States draws its financial support from the NIH, which awards billions of dollars in research grants each year. Additional research is also sponsored by companies that manufacture pharmaceuticals or devices such as ultrasound scanners, but in the case of sponsored work, there are serious questions about the freedom of researchers to report results that might seem to be adverse to the interests of their sponsors.

Underlying all medical research is the fundamental requirement of modern science: In proposing a new therapeutic regime, in claiming to have discovered or developed a new medication, the burden of proof rests with the claimant. With participants allocated at random to separate target and control groups, rigorously controlled tests must yield data that are fully reported and whose analysis should indicate a positive result well beyond the range of chance occurrence. In contrast, with "alternative medicine" (as with many other claims for discoveries that seem to contradict some firmly established conventional science), there is the strong impression that there has been a shifting of this burden of proof, with the unspoken challenge to critics: I claim X; prove me wrong. This is not the route by which modern science has arrived at its impressive, accumulated record of successes. That traditional route has been described in Chapters 4 and 5. In contrast, so persuasive are many of the supporters of "alternative medicine" that the Office of Alternative Medicine (OAM) was established by the Congress in 1992, with the legislative effort headed by Senator Tom Harkin of Iowa. OAM is within the NIH, though many of its supporters wanted it to be an independent federal agency, thus not subject to normal NIH quality controls. An oversight board was appointed for OAM, but according to its critics, it is composed only of "alternative medicine" supporters. Whatever the supervising agency for "alternative medicine," we must demand the highest standards of quality controls for the testing.

The stakes in this debate are high, yet the public is still being treated to uncritical reporting. For example, in November 1997, *Time* magazine devoted a whole page to the report of an NIH panel, with a banner headline "Acupuncture Works." The report did quote an opponent who pointed out that the panel had not invited testimony from critics, but it also quoted an unnamed researcher who claimed that savings from acupuncture treatment in stroke rehabilitation and carpal tunnel syndrome could "cut the nation's annual medical bill by $11 billion." This is a good example of the way in which public opinion is surely more influenced by headlines and by such fantasies of savings than by low-key critiques of the shaky basis of the claimed "findings."

With the new millennium approaching, what do these and other cases tell us of the extent of the general public's understanding of science and its methods? Clearly, there is room for much improvement. A major part of the remedy rests in the hands of the scientists themselves. The United States stands out in the percentage of high school graduates who go to college. Most colleges require all students to take some science courses as part of their general education. This is where we have a great opportunity to carry the message. Regrettably, the majority of the general science courses are filled merely with facts, with little or no discussion of the scientific process or method. I would hope to see at least a modest start being made. Not only should we dazzle our students with our discoveries, but we must also ask, Why do you trust me? How have the facts been demonstrated? How have these theories been confirmed? Unless we make a start in this way, we will continue to fight rearguard actions as each new pseudoscientific claim is promoted and the scientific community is attacked for its unwillingness to accept "innovation."

Notes

1. *Nature* 391 (1998): 4; *Science* 279 (1998): 35.

2. *Social Text* 14 (1996): 217.

3. *New Politics* 6, no. 2 (1997).

4. http://www.physics.nyu.edu/faculty/sokal/. See also *Studies in the History and Philosophy of Science* 28 (1997): 219–235.

5. *Chronicle of Higher Education,* July 5, 1996, A22; *Science* 272 (1996): 1864, and *Science* 273 (1996): 673; *New Yorker,* May 27, 1996.

6. The Baltimore case is the focus of a book by Daniel Kevles that will appear late in 1998.

Works Cited

Bailey, H. *A Matter of Life or Death*. New York: G. P. Putnam's Sons, 1958.

Bauer, H. H. *Beyond Velikovsky*. Urbana: University of Illinois Press, 1984.

Beyerschen, A. *Scientists Under Hitler*. New Haven: Yale University Press, 1977.

Broad, W., and N. Wade. *Betrayers of the Truth*. New York: Simon and Schuster, 1982.

Close, F. *Too Hot to Handle*. Princeton: Princeton University Press, 1991.

Condon, E. U. *Scientific Study of Unidentified Flying Objects*. New York: Bantam Books, 1969.

Druckman, D., and J. A. Swets, eds. *Enhancing Human Performance*. Washington, D.C.: National Academy Press, 1988.

Du Toit, A. L. *Our Wandering Continents*. Edinburgh: Oliver & Boyd, 1937.

Ecker, R. L. *Dictionary of Science and Creationism*. Buffalo, N.Y.: Prometheus Books, 1990.

Edge, H. L., R. L. Morris, J. H. Rush, and J. Palmer. *Foundations of Parapsychology: Exploring the Boundaries of Human Capability*. Boston: Routledge and Kegan Paul, 1986.

Englehardt, H. T., and A. L. Caplan, eds. *Scientific Controversies*. Cambridge: Cambridge University Press, 1987.

Fletcher, R. *Science, Ideology and the Media: The Cyril Burt Scandal*. New Brunswick, N.J.: Transaction Publishers, 1991.

Franklin, A. *The Neglect of Experiment*. Cambridge: Cambridge University Press, 1986.

_____. *Experiment, Right or Wrong?* Cambridge: Cambridge University Press, 1990.

_____. *The Rise and Fall of the Fifth Force*. New York: American Institute of Physics, 1993.

Franks, F. *Polywater*. Cambridge: MIT Press, 1981.

Gardner, M. *Fads and Fallacies in the Name of Science*. New York: G. P. Putnam's Sons, 1952. Reprint, New York: Dover Books, 1957.

_____. *Science—Good, Bad and Bogus*. Buffalo, N.Y.: Prometheus Books, 1981.

_____. *The New Age: Notes of a Fringe Watcher*. Buffalo, N.Y.: Prometheus Books, 1988.

Gauquelin, M. *Birthtimes*. New York: Hill and Wang, 1983.

Golubev, G. *Nikolai Vavilov*. Translated by Vadim Sternik. Moscow: MIR Publishers, 1987.

Graham, L. R. *Science and the Soviet Social Order*. Cambridge: Harvard University Press, 1990.

Gribbin, J. R., and S. H. Plagemann. *The Jupiter Effect*. New York: Random House, 1974.

_____. *The Jupiter Effect Reconsidered*. New York: Vintage Books, 1982.

Grimm, P., ed. *Philosophy of Science and the Occult*. 2d ed. Albany: State University of New York Press, 1990.

Haberer, J. *Politics and the Community of Science*. New York: Van Nostrand Reinhold, 1969.

Hansel, C.E.M. *The Search for Psychic Power.* Buffalo, N.Y.: Prometheus Books, 1989.

Hearnshaw, L. *Cyril Burt, Psychologist.* Ithaca, N.Y.: Cornell University Press, 1979.

Hines, T. *Pseudoscience and the Paranormal.* Buffalo, N.Y.: Prometheus Books, 1988.

Huizenga, J. *Cold Fusion: The Scientific Fiasco of the Century.* Rochester, N.Y.: University of Rochester Press, 1992.

Hyman, R. *The Elusive Quarry.* Buffalo, N.Y.: Prometheus Books, 1989.

Hynek, J. A. *The UFO Experience.* New York: Henry Regnery, 1972.

Jahn, R. G., and B. J. Dunne. *Margins of Reality.* San Diego: Harcourt Brace Jovanovich, 1987.

Jeffreys, H. *The Earth.* Cambridge: Cambridge University Press, 1924.

Joravsky, D. *The Lysenko Affair.* Chicago: University of Chicago Press, 1970.

Joynson, R. B. *The Burt Affair.* New York: Routledge, 1989.

Kamin, L. J. *The Science and Politics of IQ.* New York: Halsted Press, 1974.

Klass, P. J. *UFOs—Identified.* New York: Random House, 1968.

Krauss, L. M. *The Fifth Essence.* New York: Basic Books, 1989.

Kuhn, T. S. *The Structure of Scientific Revolutions.* Chicago: University of Chicago Press, 1962.

Laudan, L. *Science and Relativism.* Chicago: University of Chicago Press, 1990.

Le Grand, H. E. *Drifting Continents and Shifting Theories.* Cambridge: Cambridge University Press, 1988.

Lyttleton, R. A. *Man's View of the Universe.* Boston: Little, Brown, 1961.

Mallove, E. F. *Fire from Ice.* New York: John Wiley and Sons, 1991.

Mauskopf, S. H. *The Reception of Unconventional Science.* Boulder: Westview, 1979.

Medawar, P. B. *The Art of the Soluble.* London: Methuen, 1967.

———. *Induction and Intuition in Scientific Thought.* London: Methuen, 1969.

Medvedev, Z. *The Rise and Fall of T. D. Lysenko.* New York: Columbia University Press, 1969.

Menzel, D. H. *Elementary Manual of Radio Propagation.* Englewood Cliffs, N.J.: Prentice-Hall, 1948.

———. *Flying Saucers.* Cambridge: Cambridge University Press, 1953.

Morris, H. M., and G. E. Parker. *What Is Creation Science?* San Diego: Creation Life Publishers, 1982.

Nelkin, D. *Science Textbook Controversy and the Politics of Equal Time.* Cambridge: MIT Press, 1977.

———. *The Creation Controversy: Science of Scripture in the Schools.* New York: W. W. Norton, 1982.

Newton, I. *Principia Mathematica.* Edited by Florian Cajori. Berkeley: University of California Press, 1962.

Numbers, R. L. *The Creationists.* New York: Knopf, 1992.

Palmer, J. A., C. Honorton, and J. Utts. *Reply to the National Research Council Study on Parapsychology.* Research Triangle Park, N.C.: Parapsychological Association, 1988.

Popper, K. *Conjectures and Refutations.* New York: Basic Books, 1962.

Randi, J. *Flim-Flam! The Truth About Unicorns, Parapsychology, and Other Delusions.* New York: Lippincott and Crowell, 1980.

Raup, D. M. *The Nemesis Affair.* New York: W. W. Norton, 1986.

Sachs, M. *The UFO Encyclopedia.* New York: G. P. Putnam's Sons, 1980.

Seabrook, W. *Dr. Wood.* New York: Harcourt, Brace, 1941.

Snyderman, M., and S. Rothman. *The IQ Controversy.* New Brunswick, N.J.: Transaction Books, 1988.

Taubes, G. *Bad Science: The Short Life and Weird Times of Cold Fusion.* New York: Random House, 1993.

Taylor, J. G. *Superminds.* New York: Viking Press, 1975.

_____. *Science and the Supernatural.* New York: E. P. Dutton, 1980.

Velikovsky, I. *Worlds in Collision.* New York: Macmillan, 1950.

_____. *Earth in Upheaval.* New York: Doubleday, 1955.

_____. *Stargazers and Gravediggers.* New York: Quill, 1984.

Vitaliano, D. *Legends of the Earth.* Bloomington: Indiana University Press, 1973.

Williams, L. P. *Relativity Theory: Its Origins and Impact on Modern Thought.* New York: John Wiley and Sons, 1968.

Ziman, J. *Public Knowledge.* Cambridge: Cambridge University Press, 1968.

_____. *Reliable Knowledge.* Cambridge: Cambridge University Press, 1978.

_____. *An Introduction to Science Studies.* Cambridge: Cambridge University Press, 1984.

Further Readings

Abell, G. O., and B. Singer, eds. *Science and the Paranormal*. New York: Charles Scribner's Sons, 1981.

Alcock, J. *Science and Supernature: A Critical Appraisal of Parapsychology*. Buffalo, N.Y.: Prometheus Books, 1990.

Bauer, H. H. *The Enigma of Loch Ness*. Urbana: University of Illinois Press, 1986.

_____. *Scientific Literacy and the Myth of the Scientific Method*. Urbana: University of Illinois Press, 1992.

Beloff, J. *The Relentless Questions: Reflections on the Paranormal*. Jefferson, N.C.: McFarland and Co., 1991.

Bernal, J. D. *The Social Function of Science*. London: Routledge and Kegan Paul, 1939. Reprinted Cambridge: MIT Press, 1967.

Blinderman, C. *The Piltdown Inquest*. Buffalo, N.Y.: Prometheus Books, 1986.

Blum, H. *Out There: The Government's Secret Quest for Extraterrestrials*. New York: Simon and Schuster, 1990.

Broughton, R. S. *Parapsychology: The Controversial Science*. New York: Ballantine Books, 1991.

Butcher, H. *Human Intelligence*. London: Methuen, 1968.

Chubin, D., and E. J. Hackett. *Peerless Scinece*. Albany: State University of New York Press, 1990.

Cole, S. *Making Science*. Cambridge: Harvard University Press, 1992.

Friedlander, M. W. *The Conduct of Science*. Englewood Cliffs, N.J.: Prentice-Hall, 1972.

Futuyama, D. J. *Science on Trial: The Case for Evolution*. New York: Pantheon Books, 1983.

Geisler, N. L. *The Creator in the Courtroom: Scopes II*. Milford, Mich.: Mott Media, 1982.

Godfrey, L., ed. *Scientists Confront Creationism*. New York: W. W. Norton, 1983.

Goldsmith, D., ed. *Scientists Confront Velikovsky*. Ithaca, N.Y.: Cornell University Press, 1977.

Grove, J. W. *In Defence of Science*. Toronto: University of Toronto Press, 1983.

Hallam, A. *Great Geological Controversies*. Oxford: Oxford University Press, 1983.

Jevons, J. R. *Science Observed*. London: George Allen and Unwin, 1973.

Kitcher, P. *Abusing Science: The Case Against Creationism*. Cambridge: MIT Press, 1982.

Klass, P. J. *The Crashed Saucer Cover-Up*. Buffalo, N.Y.: Prometheus Books, 1993.

Kohn, A. *False Prophets*. Oxford: Basil Blackwell, 1986.

Lawrence, S. A. *The Battery Additive Controversy*. Tuscaloosa: University of Alabama Press, 1962.

Miller, D. J., and M. Hersen. *Research Fraud in the Behavioral and Biomedical Sciences.* New York: John Wiley and Sons, 1992.

Oberg, J. E. *UFOs and Outer Space Mysteries.* Norfolk, Va.: Donning, 1982.

Peat, F. D. *Cold Fusion: The Making of a Scientific Controversy.* Chicago: Contemporary Books, 1989.

Popovsky, M. *The Vavilov Affair.* Hamden, Conn.: Anchor Books, 1985.

Rasmussen, R. M. *The UFO Literature: A Comprehensive Annotated Bibliography of Works in English.* Jefferson, N.C.: McFarland and Company, 1985.

Ravetz, J. R. *Scientific Knowledge and Its Social Problems.* Oxford: Oxford University Press, 1971.

Rohrlich, F. *From Paradox to Reality.* Cambridge: Cambridge University Press, 1987.

Rothman, M. *A Physicist's Guide to Skepticism.* Buffalo, N.Y.: Prometheus Books, 1988.

Sagan, C., and T. Page. *UFOs: A Scientific Debate.* New York: W. W. Norton, 1974.

Schilpp, P. A. *Albert Einstein: Philosopher-Scientist.* New York: Harper and Row, 1949.

Stoddard, C. *Krebiozen.* Boston: Beacon Press, 1955.

Thomson, K. S. *Living Fossil: The Story of the Coelacanth.* New York: W. W. Norton, 1991.

Wegener, A. L. *The Origin of Continents and Oceans.* Translated by J. C. A. Skerl. London: Methuen, 1924.

Will, C. M. *Was Einstein Right?* New York: Basic Books, 1986.

About the Book and Author

Scientific discoveries are constantly in the news. Almost daily we hear about new and important breakthroughs. But sometimes it turns out that what was trumpeted as scientific truth is later discredited, or controversy may long swirl about some dramatic claim.

What is a nonscientist to believe? Many books debunk pseudoscience, and some others present only the scientific consensus on any given issue. In *At the Fringes of Science* Michael Friedlander offers a careful look at the shadowlands of science. What makes Friedlander's book especially useful is that he reviews conventional scientific method and shows how scientists examine the hard cases to determine what is science and what is pseudoscience.

Emphasizing that there is no clear line of demarcation between science and nonscience, Friedlander leads the reader through case after entertaining case, covering the favorites of "tabloid science" such as astrology and UFOs, scientific controversies such as cold fusion, and those maverick ideas that were at first rejected by science only to be embraced later.

There are many good stories here, but there is also much learning and wisdom. Students of science and interested lay readers will come away from this book with an increased understanding of what science is, how it works, and how the nonscientist should deal with science at its fringes.

Michael W. Friedlander is professor of physics at Washington University in St. Louis. He is the author of *The Conduct of Science, Astronomy—From Stonehenge to Quasars,* and *Cosmic Rays,* a selection of the Astronomy Book Club.

Index

Printed in the United States
88669LV00005B/145/A